KU-623-484

SPRINKLER IRRIGATION
equipment and practice

Melvyn Kay
BSc, MSc, C Eng, MICE

B T Batsford Limited   London

© Melvyn Kay 1983
First published 1983
Reprinted 1988

All rights reserved. No part of this publication may
be reproduced, in any form or by any means,
without permission from the Publisher.

Typeset by Tek-Art Ltd Kent
and printed in Great Britain by
Anchor Brendon Ltd
Tiptree, Essex.
for the publishers
B T Batsford Limited
4 Fitzhardinge Street
London W1H 0AH

ISBN 0 7134 12291

(m) 631.7K

SPRINKLER IRRIGATION
equipment and practice

THE COLLEGE of West Anglia

LANDBEACH ROAD • MILTON • CAMBRIDGE
TEL: (01223) 860701

24. JA.....

0 8 FEB 2005

2 8 JUN 2007

WITHDRAWN

LEARNING Centre

The ticket - holder is responsible for
the return of this book

3 8079 00054 685 3

CONTENTS

ACKNOWLEDGMENT

The author and publishers thank the following individuals and firms for supplying photographs for inclusion in this book: R Blakeman for figure 4.23; M K V Carr for figures 3.12, 6.8; J Coote for figure 2.14; Evenproducts Ltd for figures 1.2(d), 5.1, 5.2(a) (c) (d), 5.3(a) and (b); Farrow Irrigation Ltd for figures 2.5, 4.1, 4.3(a) – (c), 4.10, 4.11, 4.12, 4.13, 4.17, 4.18, 4.19, 4.20, 4.24; R Hobcraft for figures 6.1(a), 6.9(a) and (b); Nelson Irrigation Corporation for 4.6(a), 4.9(a) and (c); J D Pett and Sons Ltd for figure 6.10(b); S P P Systems Ltd for figures 6.1(a) and (b), 6.7; and Wright Rain Ltd for figures 1.2(a) and (b), 3.4(a), 3.7, 3.9, 3.17, 3.18, 3.22(a), 3.23(a) and (b), 3.25, 3.27, 3.28, 3.29, 3.31, 3.36(a) – (d), 3.39, 4.4, 4.15, 5.5, 5.8(a) and (b), 8.2(e), 8.3(a) (b), 8.4(a) (b) and (e), 8.8(c), 8.9(a) and (b).

Thanks are also due to Gordon Bennington for his help and advice on many aspects of sprinkler irrigation and for reading and commenting on the manuscript; Norman Murphy MSIAD and George Dilks for preparing artwork for the diagrams and my wife, Judith, for her encouragement during the preparation of the manuscript and for preparing the index.

PREFACE

I have written this book primarily for those students who are studying irrigation at a vocational level in developing countries. It provides practical information on a wide range of sprinkler systems from simple hand-moved equipment to the more sophisticated mobile machines. Details are given about the efficient and safe use of equipment in the field and how it should be maintained to provide long, useful service.

Also included are chapters on the basic hydraulics of sprinklers, pipes and pumps in a way that can be easily understood by those with a limited technical or mathematical background. Although I do not consider it necessary for the users of sprinkler systems to understand the complexities of their design, a working knowledge of how water is pumped, flows through pipes and is distributed by sprinklers will be invaluable to those who wish to make full and proper use of their equipment.

I have used diagrams and photographs wherever possible to illustrate equipment and operating principles to try and overcome some of the difficulties that many students have studying English texts when this is not their first language.

Although the book is aimed at the field man involved in the operation of sprinklers I hope that many of the practical aspects of the text will also be of value to engineers, agriculturalists and others involved in irrigation at both middle and professional levels.

Several chapters have in fact, formed the basis of lectures given on short courses to farmers and growers in the UK who have also been developing a keen interest in irrigation in recent years. This group too should find the book useful as the physical principles of sprinkler irrigation apply equally well in UK as in the tropics.

MK 1983

1

INTRODUCTION

Irrigation is practised in many areas of the world to provide water for crop growth. In dry areas such as the Middle East, India, Western USA and Australia there is little or no rainfall and total irrigation provides all the water needs of the crop. In humid and temperate areas such as Central Africa and Europe, although crops are grown under natural rainfall, it is often insufficient and badly distributed. Extra water is sometimes applied, supplementing rainfall, to improve yields and crop quality.

In many countries irrigation is traditionally carried out by surface flooding, using basins, borders and furrows to distribute water. In recent years, however, there has been a rapid increase in the use of sprinkler or overhead irrigation. This is a method of distributing water in pipes under pressure and spraying it into the air so that it breaks up into small water droplets and falls to the ground like natural rainfall. Sprinklers generally need less water and labour than surface irrigation and can be adapted to more sandy, erodible soils on undulating ground.

There are many types of sprinkler system available. This is because of the wide variety of soil conditions and crops to which sprinklers must be adapted. However, all the systems (figure 1.1) have the following basic components in common:

Pump

Mainline

Lateral

Sprinklers.

Pump
The pump draws water from a source, such as a reservoir, borehole or stream, and delivers it into the irrigation system. It is driven by a power unit such as an internal combustion engine or an electric motor.

Mainline
The mainline is a pipe which delivers water from the pump to the laterals (see following paragraph). In some cases the mainline is permanent and is laid in the field either above or, more usually, below ground. In others it is portable and can be moved from field to field. Permanent pipes are usually made of galvanised steel, asbestos-cement or plastic. Portable pipes are usually made of light-weight aluminium alloy, galvanised steel or plastic so that they are easily moved from place to place.

Lateral
The lateral is a pipe which delivers water from the mainline to the sprinklers. It can be portable or permanent and is made of similar materials to the mainline, but is usually smaller in size.

Sprinklers
The two main types of sprinklers used in agriculture are the rotary sprinkler and the sprayline (figure 1.2(a)-(d)).

The **rotary** sprinkler is the most commonly used type. Some systems use many small rotary sprinklers operation together (see Chapter 3 Conventional and Chapter 6

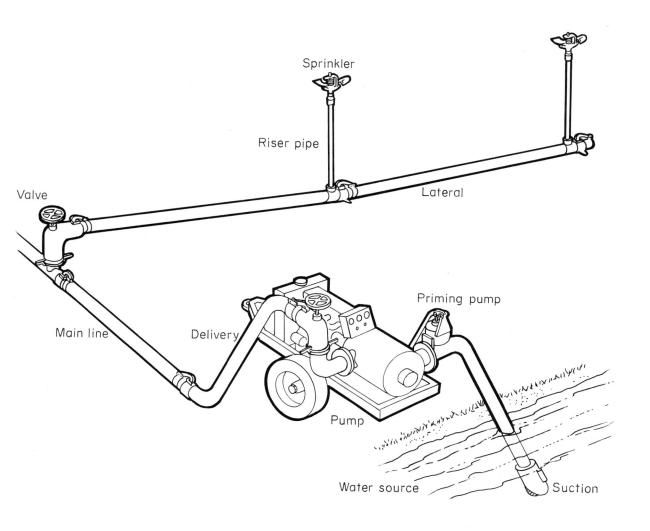

Sprinkler

Riser pipe

Lateral

Valve

Main line

Delivery

Priming pump

Pump

Water source

Suction

1.1   Components of a sprinkler system

Mobile Lateral Systems) others use only one large sprinkler or raingun (see Chapter 4 Raingun Systems).

A **sprayline** consists of a pipe with small holes or nozzles fitted along its length through which the water is sprayed. Spraylines can be stationary or oscillate from side to side, or they can be made to rotate about a central axis (see Chapter 5 Spraylines).

Many factors need to be considered when deciding which system to use. For example, some systems are more suited to undulating land or irregular shaped fields than others. Choice may also be affected by soil type and condition and the crops being grown as well as the cost of buying and maintaining the system (see Chapter 7 Choosing a Sprinkler System).

Many of the most common sprinkler systems being used are described in the following chapters including their advantages and disadvantages. Details are provided about how they work, how they can be used efficiently and safely and how they should be maintained to provide long, useful service.

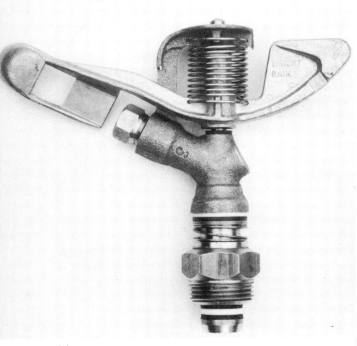

(a)

(b)

(d)

*1.2   Two main types of sprinkler (a) Rotary sprinkler
(b) Rotary sprinkler system (c) Sprayline
(d) Sprayline system*

(c)

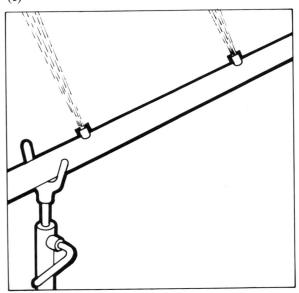

2

HYDRAULICS

2.1 Objectives

In sprinkler irrigation systems, water is pumped from a source (eg well, river or reservoir) through pipes to the sprinklers and then sprayed as uniformly as possible over the crops. The detailed design of the system is usually carried out by an experienced engineer. His job is to select the right type and size of pumps and power units, the size of pipes and the type of sprinklers. To use sprinkler equipment it is not essential to understand the complexities of its design. However, a working knowledge of how water is pumped and flows through pipes and how it is distributed by sprinklers will help the irrigator to make full and proper use of his equipment.

BASIC DEFINITIONS

2.2 SI Units (International Metric System)

SI units are slowly replacing Imperial and European units in irrigation but the latter systems are likely to be around for some time yet. Indeed a good deal of confusion exists as irrigators tend to use more than one system of units. For example, pipe sizes may be given in SI units, eg 75 mm or 100 mm diameter, and yet pump capacities may be given in Imperial units, eg 300 gallons per minute. To avoid such confusion in this text the SI system is used throughout. Equivalent units in the Imperial system, however, are quoted in parenthesis to help the reader become accustomed to the new units.

The fundamental units of the SI system are:

Measurement	Units	Symbol
Length	metre	m
Area	square metre	m^2
Volume	cubic metre	m^3
Mass	kilogramme	kg
Force	newton	N

2.3 Pressure

Pressure, in general terms, is a measure of the energy required to operate a sprinkler system. More specifically it is defined as a force acting uniformly over an area. Pressure is measured in newtons per square metre (N/m^2). This is a very small unit and a more practical one is the kilonewton per square metre (kN/m^2). An alternative to this is the **bar**. 1 bar is equal to $100 \, kN/m^2$ and is approximately 14.5 $lbf/in.^2$). A typical operating pressure for a small rotary sprinkler system is $300 \, kN/m^2$ or 3 bar (44 $lbf/in.^2$). This means that every square metre of the inside surface of the pipes has a uniform force of 300 kN acting on it.

Other common units of pressure are pounds force per square inch ($lbf/in.^2$) Imperial units, and kilogramme force per square centimetre (kgf/cm^2) European units.

Pressure measurement

Pressure in a pipe system can be measured using a Bourdon gauge (figure 2.1). Inside the gauge is a curved tube of oval section which tries to straighten out when the system is under pressure. The tube is linked to a pointer which moves across a graduated scale and records the pressure.

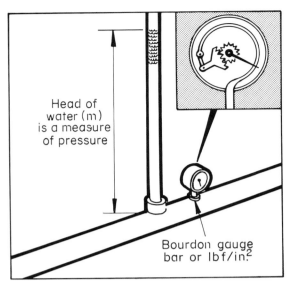

2.1 Measuring pressure

Irrigators normally measure pressure in the field using these gauges as they are robust and simple to use. Frequently, however, design engineers refer to pressure as a *head of water* as this is often more convenient for their use. If the Bourdon gauge is replaced by a long vertical tube the water pressure in the pipe would cause water to rise up the tube (figure 2.1). The height of this water column is a measure of the pressure in the pipe. For example, a pressure of 3 bar on the Bourdon gauge would result in water rising to a height of 30 m in the tube. It is simple to change from pressure to head of water:

In SI units:
$$\text{Head of water (m)} = 0.1 \times \text{Pressure (kN/m}^2)$$
$$\text{or} = 10 \times \text{Pressure (bar)}$$

In Imperial units:

$$\text{Head of water (ft)} = 2.31 \times \text{Pressure (lbf/in}^2.)$$

An example in the Imperial system shows that a pressure of 45 lbf/in.^{2} is equivalent to 104 ft head of water.

2.4 Discharge
The speed with which water flows in a pipe is called the *velocity*. which is measured in metres per second (m/s). The *discharge* is the volume of water flowing along the pipe each second which is measured in cubic metres per second (m^3/s). To understand this, consider the case of water flowing in a 100 mm diameter pipe (cross sectional area 0.008 m^2) at 1.5 m/s (figure 2.2). In one second the quantity of water flowing past ×–× will be the shaded volume. The length of the shaded portion is numerically equal to the velocity, in this case 1.5 m. The volume flowing each second or discharge is equal to the pipe area multiplied by the shaded length,
ie $1.5 \times 0.008 = 0.012$ m^3/s.

In more general terms we can write:

Discharge (m^3/s) = cross sectional area of pipe (m^2)
$\times$ velocity of flow (m/s)

For small sprinkler systems this unit of discharge is too small. An alternative is to use cubic metres per hour (m^3/h). In the example given 0.012 m^3/s is equal to 43.2 m^3/h. The conversion is made by multiplying by 3600.

Discharge measurement
Discharge in a pipeline can be measured using a discharge or flow meter (figure 2.3(a)). The meter indicates the volume of water passing through the pipeline. By noting time taken to do this the discharge can be found from:

$$\text{Discharge (m}^3/\text{s)} = \frac{\text{Volume of water (m}^3)}{\text{Time taken (s)}}$$

Discharge from a rotary sprinkler can be measured simply by connecting a flexible tube to the sprinkler nozzle and collecting a known volume of water in a container over a specified period (figure 2.3(b)). The discharge can then be found using the above formula.

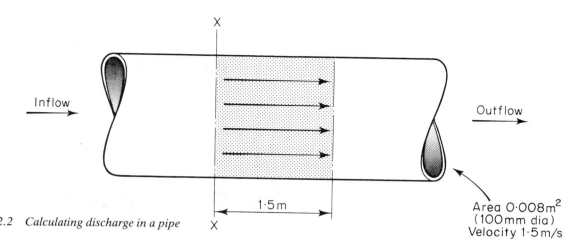

Inflow →

Outflow →

1·5 m

Area 0·008m²
(100mm dia)
Velocity 1·5 m/s

2.2   Calculating discharge in a pipe

Example

A small plastic tube is connected to a sprinkler nozzle and the discharge is collected in a bucket. The bucket can hold 5 litres of water and it takes 15 seconds to fill. What is the sprinkler discharge?

$$\text{Volume of bucket} = 5 \text{ litres}$$
$$= 0.005 \text{ m}^3$$

$$\text{Discharge (m}^3\text{/s)} = \frac{\text{Volume (m}^3)}{\text{time taken (s)}}$$

$$= \frac{0.005}{15}$$

$$= 0.00033 \text{ m}^3\text{/s}$$

This is a very small discharge. An alternative is to use m³/h.

$$\text{Discharge (m}^3\text{/h)} = \text{Discharge (m}^3\text{/s)} \times 3600$$
$$= 0.00033 \times 3600$$
$$= 1.2 \text{ m}^3\text{/h}$$

2.3   Measuring discharge (a) Flow meter used for pipelines (b) From a single sprinkler

ROTARY SPRINKLERS

2.5 Introduction

Rotary sprinklers (figure 2.4) are the most important component in an irrigation system as their performance determines the effectiveness and efficiency of the whole system.

A sprinkler operates by forcing water under pressure through a small circular hole or nozzle and into the air. The jet gradually breaks up into small drops as it travels through the air and falls to the ground like natural rainfall. A continuous curtain of water drops is distributed over the ground surface (figure 2.5). The sprinkler rotates in a horizontal direction and produces a circular wetting pattern (figure 2.6). The distance from the sprinkler to the outer edge of the circle is called the *throw*. A typical small rotary sprinkler will irrigate a circle of about 36 m diameter.

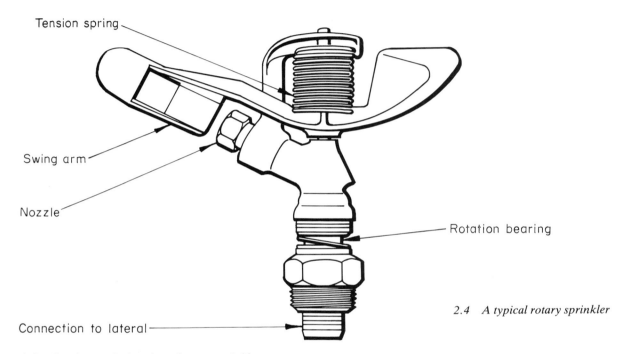

Tension spring

Swing arm

Nozzle

Rotation bearing

Connection to lateral

2.4   *A typical rotary sprinkler*

2.5   *Good water jet break up from a sprinkler*

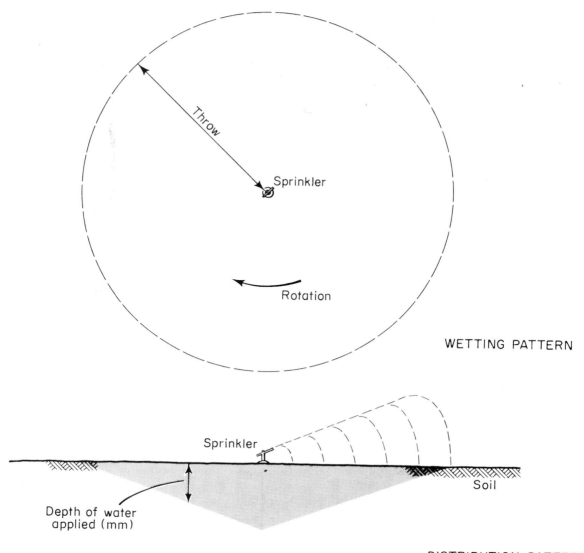

Throw

Sprinkler

Rotation

WETTING PATTERN

Sprinkler

Soil

Depth of water
applied (mm)

DISTRIBUTION PATTERN

2.6   *Wetting and distribution patterns from one
sprinkler*

Method of rotation

Sprinkler rotation is caused by the water jet and the spring-loaded swing arm. When the sprinkler is operating, the swing arm interrupts the water jet and is forced sideways by the flow (figure 2.7). Once clear of the jet the arm returns, owing to the spring tension, and interrupts the jet again. On returning however, the arm strikes one side of the sprinkler causing it to turn slightly. This action is repeated in a steady beating motion causing the sprinkler to rotate slowly. The speed is controlled by the swing arm spring tension. It is important that the sprinkler rotates at the right speed so that no area is left unirrigated (figure 2.8).

Rotary sprinklers when working satisfactorily should:

   Distribute water evenly
   Control the water application rate
   Produce an acceptable range of drop sizes.

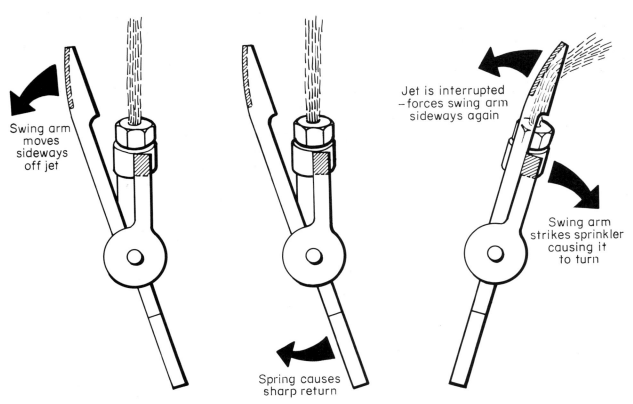

Swing arm moves sideways off jet

Spring causes sharp return

Jet is interrupted – forces swing arm sideways again

Swing arm strikes sprinkler causing it to turn

2.7   *Method of sprinkler rotation*

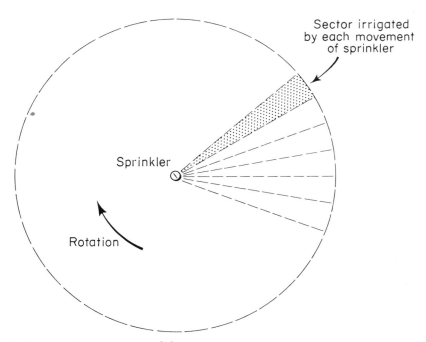

Sector irrigated by each movement of sprinkler

Sprinkler

Rotation

2.8   *Speed of rotation is chosen so that no areas are left unirrigated*

2.6 Water distribution

It is difficult to design a rotary sprinkler which produces an even irrigation over the whole of the wetted circle. Normally the application is heaviest close to the sprinkler and reduces towards the edge. The pattern of distribution is shaped like a triangle (figure 2.6). To make the distribution more even or uniform, several sprinklers are operated close together so that their distribution patterns overlap (figure 2.9). This determines the spacing needed between sprinklers. For good uniformity overlap should be 65% of the wetted diameter. Uniformity can be improved by putting them much closer together but this may lead to problems of high water application rates. The number of sprinklers used also increases, raising the cost of the system.

The uniformity of distribution from a stationary sprinkler system can be tested in the field. To do this several small cans are placed in a square grid between the sprinklers (figure 2.10). The system is then operated for a typical irrigation set time and water is collected in the cans. By measuring the depth of water in each of the cans it is possible to see just how uniform is the irrigation.

Uniformity of distribution for a mobile system can be tested by setting a line of cans across the travel path of the machine.

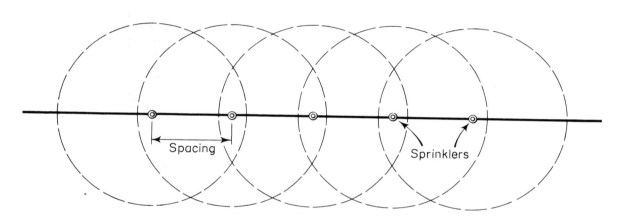

WETTING PATTERN

INDIVIDUAL SPRINKLER PATTERNS OVERLAPPED

RESULTING DISTRIBUTION OF WATER IN SOIL

2.9 Wetting and distribution patterns from several sprinklers operating close together

17

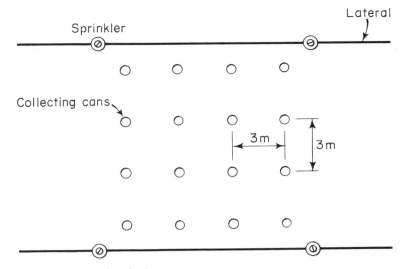

2.10   Layout of cans to test sprinkler distribution pattern

2.7 Application rate

The rate at which sprinklers apply water when a group of them is operating close together is called the *application rate*. This is measured in millimetres per hour (mm/h). In Imperial units this would be inches per hour (in./h).

The application rate depends on the size of the sprinkler nozzle, the operating pressure and the spacing between the sprinklers. Increasing the nozzle size or pressure and bringing the sprinklers closer together will increase the application rate. Manufacturers normally supply such information about their sprinklers for the design engineer to use. The application rate should always be less than the rate at which the soil can absorb water. This will avoid surface water run-off and possible soil erosion. Table 2.1 provides a guide to typical nozzle sizes, sprinkler spacings and application rates.

2.8 Drop sizes

A sprinkler normally produces a wide range of drop sizes from 0.5 mm up to 4.0 mm in diameter. Drops larger than this tend to break up into smaller drops. The smaller drops usually fall close to the sprinkler while the larger ones travel much further.

Large drops can damage delicate crops and some soils by breaking down the surface structure and reducing the infiltration rate (a process known as soil capping). In such cases, only sprinklers producing small drops should be used to lessen the damage.

The range of drop sizes can be controlled by the size of nozzle and its operating pressure. Table 2.2 provides a guide to the nozzle sizes used for sprinklers and the pressures required to break up the water jet properly in to drops. At lower pressures drops tend to be large. At higher pressures they are much smaller and misting may occur.

Table 2.1   Typical sprinkler data

Nozzle diameter (mm)	Pressure (Bar)	Diameter of wetted circle (m)	Flow m³/h.	Application rate mm/h. for spacings (m)		
				18×18	18×24	24×24
4	3.0	29	1.02	3.2	–	–
5	3.0	32	1.67	5.2	3.8	–
6	3.0	35	2.44	7.5	5.7	4.2
8	4.0	43	4.96	15.3	11.4	8.6
10	4.5	48	8.13	25.1	18.9	14.0

Nozzle size (mm)	Adequate jet break up bar (lbf/in.²)*	Preferred pressure range bar	(lbf/in.²)*
3.0 to 4.5	2.00 (30)	2.75 to 3.50	(40-50)
4.5 to 6.0	2.75 (40)	3.50 to 4.25	(50-60)
6.0 to 19.0	3.50 (50)	4.25 to 5.00	(60-70)

*Imperial units are approximate values only.

Table 2.2 A guide to nozzle sizes and pressure for proper break up of water jets

Wind speed (m/s)	Diameter of wetted circle (m) 32	37	42
	Sprinkler spacing (m)		
no wind	21	24	27
0–2.5	18	21	24
2.5–5.0	15	18	21
over 5.0	9	12	12

Table 2.3 Effect of wind speed on sprinkler spacing

2.9 Factors affecting performance

The evenness of water distribution can be seriously affected by wind and operating pressure.

Wind

Spray from sprinklers is easily blown by wind and this can distort wetting patterns and upset irrigation uniformity (figure 2.11). To reduce the effects of wind, the sprinklers can be brought closer together. The effects of different wind speeds on the required spacing of sprinklers is shown in the table 2.3.

Although 5 m/s is only thought of as a gentle breeze, it will seriously disrupt the operation of a sprinkler system. Sprinklers need to operate very close together under these conditions to distribute water evenly. In prevailing wind conditions the designer will normally position the lateral at right angles to the wind direction and reduce the sprinkler spacing along the lateral.

Operating pressure

A sprinkler performs best at a given pressure which is normally specified by the manufacturer. If the pressure is substantially above or below this recommended value then the distribution of water can be quite different from that normally expected (figure 2.12).

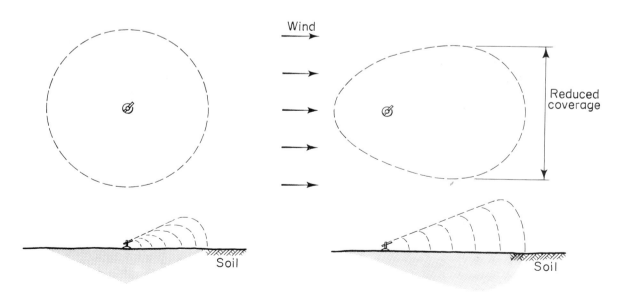

NO WIND

WINDY CONDITIONS

2.11 Effect of wind on sprinkler performance

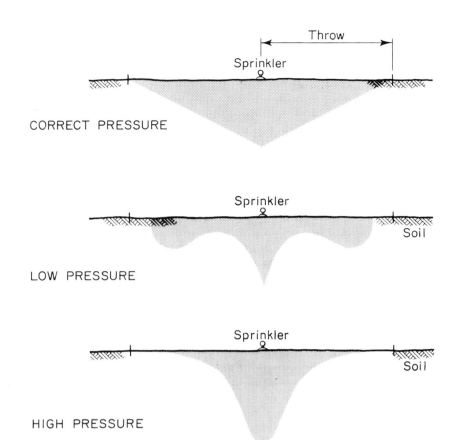

CORRECT PRESSURE

LOW PRESSURE

HIGH PRESSURE

2.12   Effect of operating pressure on sprinkler performance

If the pressure is too low the water jet does not break up easily and most of the water falls some distance from the sprinkler. Large water drops are produced and these can destroy soil structure and damage delicate crops. If the pressure is too high, the jet breaks up too much causing misting and most of the water falls close to the sprinkler. In each case the throw is reduced. Both these patterns are quite different from the normal triangular distribution and it is obvious that patterns such as these will not produce a uniform irrigation. The student can examine this problem for himself using a simple hand-held hose pipe (figure 2.13). When the hose outlet is open the jet does not break up and tends to fall in one place away from the hose (low pressure case). When a finger is placed over the end of the pipe this increases the pressure and the jet breaks up into a fine mist which falls much closer to the hose (high pressure case).

Operating at too low a pressure is a common problem on many sprinkler schemes. An example of this is shown in figure 2.14. Note the 'doughnut ring' effect produced in the crop caused by the poor distribution of water.

Bourdon pressure gauges on the mainline and laterals help the irrigator to maintain correct pressures in the system. However, a simple method of checking the operating pressure at a sprinkler is to use a Bourdon gauge with a pitot tube attached. This can be pushed into the sprinkler nozzle and provides an instant reading of pressure (figure 2.15).

An approximate method of checking pressure is to observe the shape of the water jet (figure 2.16). If the line of the jet is straight the sprinkler is working at the correct pressure. If it is bow shaped then the pressure is too low and should be increased.

2.10 Set time

The word *set* is one which is commonly used in irrigation. It refers to an area irrigated by a sprinkler or a group of sprinklers. The *set-time*

Low pressure

High pressure

2.13   *Effects of pressure on the breakup of water jet from a hose pipe*

is the time taken for sprinklers to complete an irrigation in one position.

The set-time depends on the sprinkler application rate and the irrigation need.

Example

A sprinkler system applies water at 10 mm/h to a field requiring a 90 mm irrigation. What is the set-time?

$$\text{Set time} = \frac{\text{Irrigation need}}{\text{Application rate}}$$
$$= \frac{90}{10}$$
$$= 9 \text{ hours.}$$

2.14   *Aerial view of a sugar estate showing effect of poor water distribution*

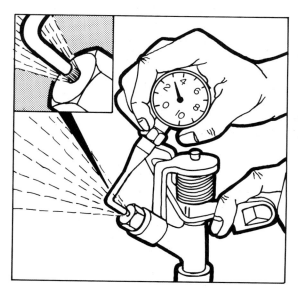

2.15   *Using a Bourbon gauge with a pilot attachment to measure the sprinkler operating pressure*

21

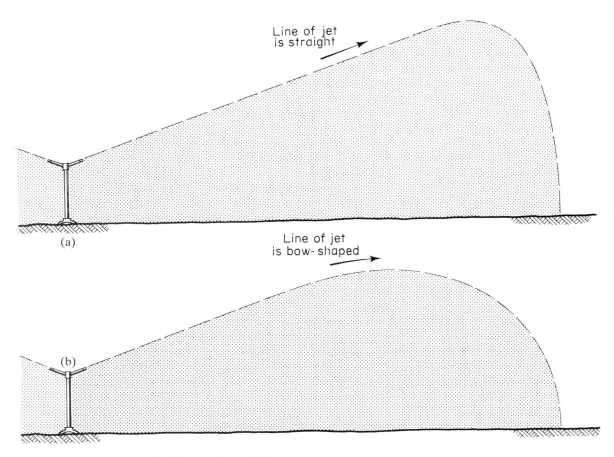

2.16 An approximate method of assessing sprinkler operating pressure. (a) Correct pressure (b) Pressure too low

If the irrigation need is only 60 mm during the early part of the season, the set time will be $^{60}/_{10} = 6$ hours.

Once an irrigation system is installed, changes in the amount of water applied can only be made by varying the set time. It is not possible to change the application rate as this is fixed by the sprinklers, pipe system and pump. Any attempt to change the application rate by changing the pressure will upset the evenness of water distribution.

FLOW IN PIPES

2.11 Introduction
Pipes are used to supply water to the sprinklers. Their size, wall thickness and strength depend on the discharge they must carry and the pressures required in the system. It is difficult to design a system that will provide the right pressure at every sprinkler. Normally, pressures vary throughout a pipe system as losses occur from friction. Pressures are usually highest at the head of a system, close to the pump, and gradually reduce towards the tail.

2.12 Pipe sizes and safe pressures
As sprinkler systems work under pressure, pipes must be able to withstand high pressures without bursting. Most countries now specify standards to which pipes should be made. They are tested to very high pressures before leaving the factory so that they will easily stand up to the much lower pressures used in practice.

Pipes are usually specified by their internal diameter or external diameter depending on the material from which they are made.

Portable irrigation pipes however usually have thin walls and are known by their nominal size and the difference between internal and external diameter is ignored.

2.13 Factors affecting pipe flow

As water flows along a pipe there is a gradual loss in pressure from friction. Although the inside of pipes may seem very smooth to touch, they can be quite rough hydraulically. This roughness slows down the water flow in the same way as friction slows down an object as it is pushed across some rough surface (figure 2.17). Pipes generally tend to increase in roughness with age. Portable aluminium pipes, for example, are easily scored by grit in the irrigation water and dented with use. Steel pipes may gradually become rusty and pitted. Chemical scales and bacterial slimes may also adhere to the pipe walls. These gradually increase the pipe roughness, can reduce the pipe diameter, and can lead to pressure losses greater than those anticipated when the pipes were new. The engineer will normally take account of these factors when designing a new system.

Pressure losses depend not only on the roughness but also on the discharge, the pipe diameter and the pipe length. If the discharge in the pipe increases, the flow velocity also increases and this causes the friction to rise very rapidly resulting in much greater pressure loss. To overcome this a larger diameter pipe can be used which has a greater discharge capacity and a much lower velocity of flow.

The length of the pipe has a direct effect on the pressure loss. The further the water has to travel the more friction it will encounter and the greater will be the pressure loss.

The relationships between these factors are complex and it is the design engineer's task to understand and use them in choosing the right size of pipe to use. The following example shows the effect each factor has on the loss of pressure.

Example
A pipeline is 1000 m long and carries a discharge of 25 m^3/h.

What effect will the pipe diameter have on the pressure loss?

Pipe diameter (mm)	75	100	125
Pressure loss (bar)	5.4	0.7	0.25

What effect will increasing the discharge to 50 m^3/h have on the pressure loss?

Pressure loss (bar)	20	2.2	0.5

What effect will increasing the pipe length to 2000 m have on the pressure loss. (assume flow is 50 m^3/h)

Pressure loss (bar)	40	4.4	1.0

To summarise:
Pressure losses are much higher in smaller diameter pipes carrying the same flow.

Pressure losses increase very rapidly as the flow increases, particularly in the small diameter pipes.

Pressure losses increase directly with pipe length. If the length doubles then the pressure loss also doubles.

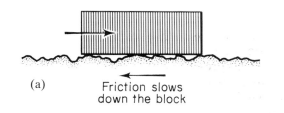

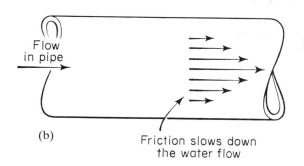

2.17 Friction slows movement. (a) Block on rough surface (b) Flow in a pipe

Table 2.3 is a guide to selecting suitable pipe sizes for different flows and pipe lengths. This is only a guide and should not be used for design as each manufacturer's pipe varies to some extent in its friction loss characteristics.

Discharge (m³/h)	Pipe length m		
	0-250	250-500	500
		Pipe diameters(mm)	
5	50	50	
10	75	75	
25	75	75	
50	100	100	
60	100	125	150
70	100	125	150
80	125	150	150

(Velocity of flow should always be less than 1.5mls).

Table 2.3   A guide to selecting pipe sizes

2.14 Effects of ground elevation on pressure

Sprinkler systems are often used in areas where the land topography is undulating or sloping steeply. Changes in ground elevation will cause changes in pressure in a pipe. For example, if a sprinkler lateral is laid on an uphill slope (figure 2.18) the pressure in the pipe will drop by 0.1 bar (1.5 lbf/in.²) for every 1.0 m rise in ground level. This pressure drop will obviously affect the sprinkler performance further up the slope, particularly

as pressure is also being lost along the pipe through friction. To avoid this problem sprinkler laterals should be laid out level along the ground contour. If this is not possible then allowances for the elevation change must be made when determining the pressures required in the system.

Laying sprinkler laterals on a gentle downhill slope can be of benefit as the pressure increases by 0.1 bar (1.5 lbf/in.²) for every 1.0 m fall in the ground elevation. This increase can be used to offset losses that occur from pipe friction. Too much fall however can cause difficulties as the pressure rise may become too high resulting in sprinklers operating above their recommended pressures.

2.15 Pumping pressure requirements

The pressure to be supplied to the sprinkler system by pumping (figure 2.19) must take account of:

Recommended pressure at sprinkler

Pressure losses in the mainline and laterals

Changes in ground elevation.

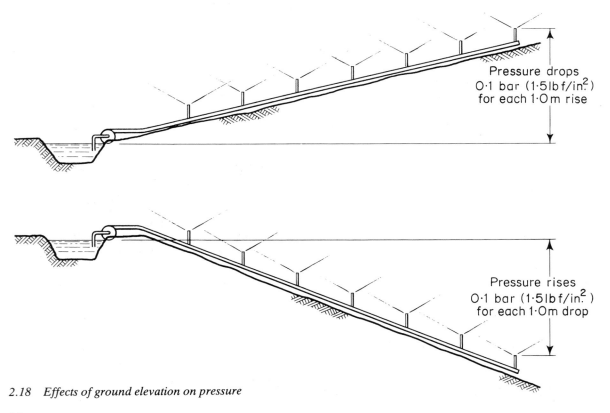

Pressure drops 0·1 bar (1·5 lbf/in.²) for each 1·0 m rise

Pressure rises 0·1 bar (1·5 lbf/in.²) for each 1·0 m drop

2.18   Effects of ground elevation on pressure

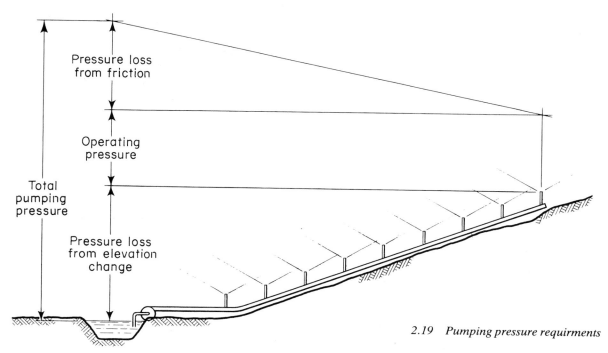

2.19 *Pumping pressure requirments*

A common fault in sprinkler systems is to install pipes which are too small. This is often done for the simple reason that small diameter pipes are cheaper than larger ones. Little attention is given to the high pressure losses that occur and faults are only realised when the sprinklers have insufficient pressure to work properly. Smaller diameter pipes also leave no room for future expansion of a system and further problems may occur as pipe roughness increases with age.

Pressure can only be maintained in the pipe system by using larger pumps to cope with the greatly increased pressure losses. This not only involves the cost of buying new pumps but also extra fuel costs in maintaining the higher pumping pressures.

2.16 Waterhammer

Waterhammer can be a serious problem in sprinkler systems. Most students will be familiar with waterhammer in a domestic context. It is the rhythmic banging noise which often occurs when a cold water tap is closed rapidly. The banging is a series of high pressure waves which move very rapidly along the pipe as the water is brought to rest at the tap. In larger pipes, such as those used in irrigation, rapid closure of pumps and valves can cause pressure rises in the pipe many times

greater than the normal working pressure. In some cases, the pressure rise may be sufficient to burst pipes and pumps.

Waterhammer is the result of sudden changes which take place in a pipe system. Examples of these are:

Starting and stopping pumps – electric pumps can close down rapidly when there is a power failure.
Rapid closure of a valve or hydrant.
Vehicle driving across a flexible hose.
Sudden blockage in a pipe or sprinkler nozzle.

Waterhammer can easily be prevented by avoiding sudden changes taking place in the flow, for example:

Start and stop pumps slowly.
Close valves or hydrants slowly.
Avoid driving over flexible hoses by using pipe bridges.
Avoid blockage problems by using filters but make sure they are cleaned regularly.

Some waterhammer is unavoidable in sprinkler systems and so it is common practice to protect the pump from damage by using a reflux valve (see 8.4 Pump delivery) page 0.

3

CONVENTIONAL SYSTEMS

3.1 Introduction

Irrigation systems using many small rotary sprinklers operating together were the first to make sprinkler irrigation popular in the 1930s, and they are still the most commonly used system today. The sprinklers operate at low to medium pressures of 2-4 bar (30-60 lbf/in.^{2}) and can irrigate an area 9-24 m wide and up to 300 m long at one setting. Application rates vary from 5-35 mm/h.

There are three main types of system in use:

Portable systems

Solid or permanent systems

Semi-permanent systems

3.2 Portable systems

Hand-moved system

The simplest portable system is designed to be moved by hand (figure 3.1). It consists of a pump, mainline, lateral and rotary sprinklers spaced 9-24 m apart. The lateral is usually between 50 mm and 125 mm in diameter so that it can be moved easily. It remains in position until irrigation is complete. The pump is then stopped and the lateral disconnected from the mainline and allowed to drain. It is dismantled and moved by hand labour to the next point on the mainline and re-assembled. As the lateral is connected to

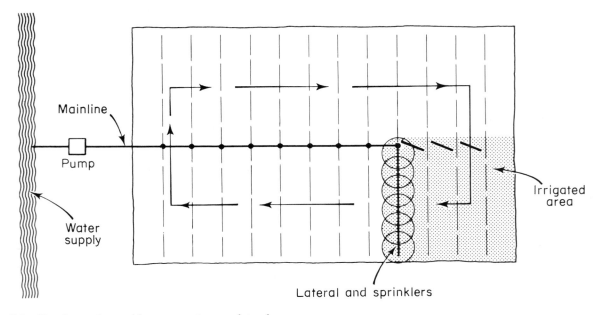

3.1   Handmoved portable system using one lateral

the end of the mainline it is also necessary to disconnect sections of the mainline. Usually, the lateral is moved between one and four times each day depending on the set time. It is gradually moved around the field until the whole field is irrigated.

Many systems use two or more laterals to irrigate larger areas (figure 3.2). They are connected to the mainline using valve couplers, see page 40. This allows irrigation to continue while one of the laterals is being moved.

In the systems described, only the laterals are moved during the irrigation season while the mainline remains permanently in place. In some cases, however, the whole system including pump and mainline is moved from field to field. Systems where part or all of the equipment is moved regularly are called portable or portable-set systems.

There are many other ways of laying out the pipe system. Figure 3.3 shows an alternative layout with the mainline down one side of the field.

Portable sprinklers are one of the most popular systems and are used to irrigate a wide range of field and orchard crops. Their capital cost is low and they are very simple to use. However, equipment must be moved regularly and this requires a large labour force often working in wet, muddy and uncomfortable conditions. Skilled operators are also needed to run and maintain the system properly.

Labour-saving systems

In areas where labour is difficult to find or is expensive it may be difficult to use simple hand-move equipment. For this reason many portable systems have been developed which reduce the labour required to move the equipment in the field.

One system uses an aluminium or galvanised steel lateral as the axle of a wheel, some 1.5-2.0 m in diameter (figure 3.4(a)). It is sometimes called a *lateral-move* or *roll-move system*. The wheels are spaced 9-12 m apart and allow the lateral to be rolled from one irrigation setting to the next. An internal combustion engine is normally used to move the system (figure 3.4(b)). It is connected to one end of the lateral through a gear box and turns the wheel using the lateral as a drive shaft. The pipes must be strong, and rigid couplings are used to carry the heavy torque loads. On some systems the engine is located in the middle of the lateral to reduce the torque.

Small rotary sprinklers are used 9-24 m apart. On smooth level land the lateral can be

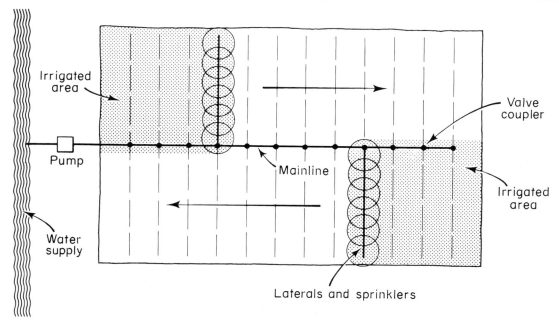

3.2   *Hand-moved portable system using two laterals*

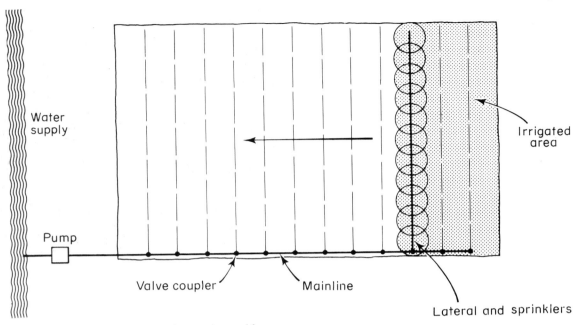

Water supply

Pump

Valve coupler

Mainline

Irrigated area

Lateral and sprinklers

3.3 Alternative layout for a hand-moved portable system

up to 500 m long. A typical layout for this system is similar to that of the alternative hand-moved one shown in figure 3.3. The mainline is laid along the side of the field. When irrigating, the lateral remains in one place until the water has been applied. The pump is stopped and the lateral uncoupled from the mainline and allowed to drain. It is then light enough to be moved to the next position using the motor. The lateral is recoupled to the mainline and irrigation restarted. To make sure that the sprinklers are always in an upright position after each move they are mounted on a special swivel assembley (figure 3.4(c)).

After the field has been irrigated the lateral is moved back to the original starting position. This means rolling the wheels over recently irrigated land. Sandy soils tend to compact when wet and are suitable for this purpose.

(a)

3.4 Lateral-move system (a) General view (b) I/C engine to move the lateral (c) Swivel assembley to ensure sprinklers always upright

(c)

(b)

Clay soils are soft and unstable and wheels may sink or become heavily caked with mud. It may be necessary to wait several days until the soil surface dries out. An alternative, though a rather tedious one, is to dismantle the system and carry it back after each irrigation.

This system is best suited to large flat rectangular areas growing low field and row crops. On row crops the lateral is placed across the crop so that the wheels will roll along the rows.

Another labour saving system is the *flexible lateral system* which can be wound up onto a drum at the end of each irrigation (figure 3.5(a)). The rotary sprinklers are connected to the lateral at intervals on special frames. They lie flat when being coiled and pop up vertically when irrigating.

3.5 Flexible lateral system.
(a) Lateral is wound onto a drum

(b) Sprinklers pop up vertically when irrigating

3.3 Solid-set or permanent systems

When sufficient laterals and sprinklers are provided to cover the whole irrigated area, so that no equipment needs to be moved, the system is called a *solid-set* system (figure 3.6).

For annual crops pipework and sprinklers are laid out just after planting and remain in the field throughout the irrigation season. Just before harvesting the equipment is taken up and stored until the next season.

When irrigating perennial crops, such as orchards, pipework and sprinklers are often left in a place from season to season. In this case the system is called a *permanent* one. Many permanent systems are buried below ground to avoid damage from farm vehicles. Occasionally they are laid out on the posts over the top of the crop (figure 3.7).

Most solid and permanent systems have only part of the system irrigating at one time. This depends on the size of the pipes and the amount of water available. Flow is diverted from one part of the system to another by hydrants or valves (see page 39). For special conditions, such as crop cooling or frost protection however, it is essential to operate the whole of the system at the same time.

Solid and permanent systems require far less labour than portable systems and large areas can be irrigated using only a few skilled operators. They are more expensive initially because of the extra pipes, sprinklers and fittings required, but savings can be made because of the reduced labour costs. Irrigation equipment like this is particularly suited to automation and is useful in areas where labour is difficult to obtain or is very expensive.

3.4 Semi-permanent systems

Many new irrigation systems have been developed in recent years with the advantages of both portable and solid-set equipment. They have tried to combine both low capital costs and low labour requirements. These are often referred to as semi-permanent systems and three of the most commonly used are:

Sprinkler-hop systems

Pipe-grid systems

Hose-pull systems

Sprinkler-hop systems

Sprinkler-hop systems are similar in many ways to portable systems but sprinklers are placed only at alternate positions along the lateral (figure 3.8). When sufficient water has been applied the sprinklers are disconnected and moved or 'hopped' along to the next position where they irrigate for a similar

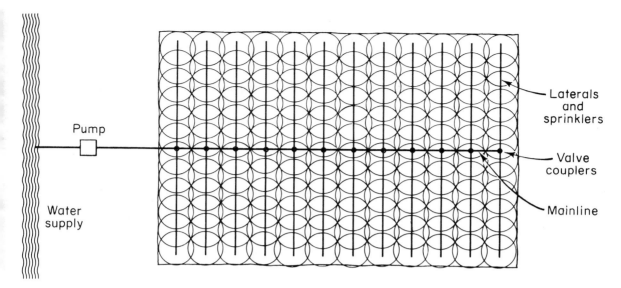

Labels on diagram:
- Laterals and sprinklers
- Valve couplers
- Mainline
- Pump
- Water supply

3.6 Solid-set or permanent system

3.7 Permanent system laid out on posts over top of crop

period (figure 3.9). This is done without stopping the flow in the lateral. Each sprinkler connection is fitted with a special valve which automatically stops the flow when the sprinkler is removed, see page 42. The lateral is then moved to the next position and the hopping process repeated.

Hop systems apply water at low rates so that only small pipes sizes and pumps are needed. Normally only one lateral move and one sprinkler hop are required each day and this task can often be conveniently fitted in with other labour work around the farm. As the system can operate for long periods un-

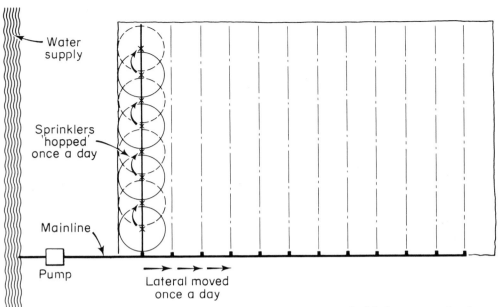

3.8   Sprinkler – hop system

3.9   Sprinklers are move or 'hopped' along to the next position along the lateral

attended it is most suited to conditions where irrigation needs to continue throughout the night.

A typical schedule for a hop system applying 70 mm of water would be:

Time	Activity
7.00 am	Labour arrives. Layout lateral in field and fix sprinklers at alternate positions.
8.00 am	Start irrigating.
6.00 pm	Labour returns. 'Hop' sprinklers to alternate positions along lateral without stopping the pump.
4.00 am	Stop irrigating by shutting down the pump. This can be done manually or automatically.
7.00 am	Labour arrives. Move lateral and sprinklers to next position.
8.00 am	Start irrigating again.

Having sprinklers at alternate positions along the lateral is only one way of using a hop system. An alternative is to have one sprinkler every three positions, giving one initial position and two hops.

Pipe-grid systems
Pipe-grid systems are similar in many aspects to solid-set systems (figure 3.10). Small diameter laterals about 25 mm diameter are used to keep system costs low. The pipes are

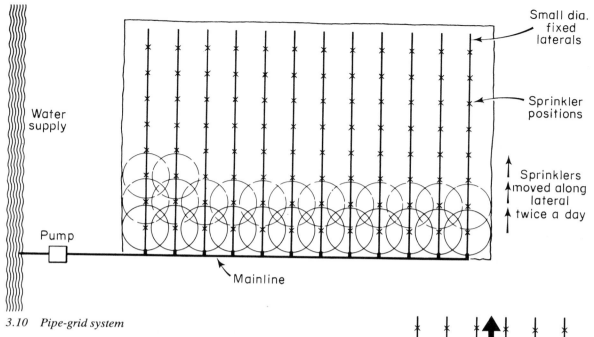

3.10 Pipe-grid system

laid out over the whole field and they remain in place throughout the irrigation season, thus eliminating pipe movement between irrigations. Two sprinklers are connected to each lateral, one near the top, the other half way down. When sufficient water has been applied each sprinkler is disconnected and moved along the lateral to the next position. This procedure is repeated until the whole field has been irrigated. The sprinklers are then returned to the beginning to repeat the operation. They are connected to laterals using valves similar to those used in the hop system.

Pipe-grid systems apply water at low rates over long periods, often throughout the night, needing very little attention. As with the hop system, the movement of sprinklers can be arranged to fit in with other farming activities. A typical system would involve at least two sprinkler moves on every lateral each day. For example, one move in the morning and one move in the evening.

A good deal of time can be saved with this system if an extra sprinkler is available and the following sequence of changing the sprinklers is adopted (figure 3.10). First place the extra sprinkler at A. Then move sprinklers B and C to D and E following the path shown.

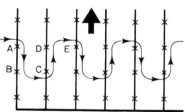

Continue across the field in this manner. The remaining sprinkler is used to begin the sequence for the next sprinkler move.

Hose-pull systems
Hose-pull systems were originally produced for citrus undertree irrigation but are now being used in other orchard crops and for some row crops (figure 3.11). The mainline and laterals are usually permanently installed either on or below the ground surface. Small diameter plastic hoses supply water from the lateral to one or two rotary sprinklers. The hose length is normally restricted to about 50 m because of friction losses in the pipe. Figure 3.11 shows the sprinkler positions for a 10-day irrigation cycle. When irrigating, the two sprinklers are placed between the tree rows in positions 1-1 and may remain there for the whole day. The next day they are pulled along to position 2-2 and so on until irrigation is complete (figure 3.12). The use of hoses in this way reduces the number of permanent laterals

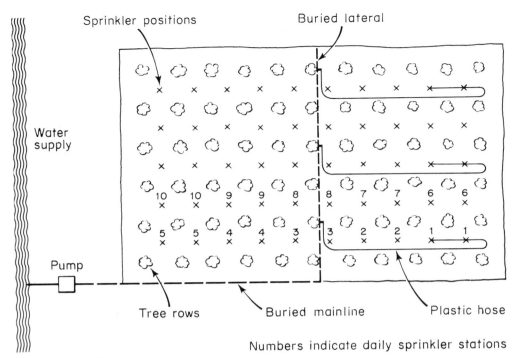

Sprinkler positions · Buried lateral · Water supply · Pump · Tree rows · Buried mainline · Plastic hose

Numbers indicate daily sprinkler stations

3.11   Hose-pull system

3.12   Pulling hose and sprinkler to new irrigation position

that are needed. They also provide great flexibility in irrigation. Sprinklers can be moved near to young trees to avoid unnecessary wetting of large areas of land. They can also be positioned easily to compensate for the distortion of spray patterns caused by wind.

Although this system normally costs less than a permanent one, problems can arise with the plastic hoses. They are easily damaged by farm implements and rough handling, and they gradually deteriorate in hot dry climates when exposed to too much bright sun light.

SYSTEM COMPONENTS

A sprinkler system is made of many different components.

3.5 Mainline and laterals

Mainlines and laterals are either permanent or portable. Permanent pipes are laid above or below ground and are generally made of galvanised steel, asbestos cement or plastic. Underground pipes do not use up land space or interfere with farming operations. Normally they are buried with a minimum cover of 0.75 m. (figure 3.13).

Portable pipes which are moved frequently need to be strong and yet light and easy to handle. They are usually made of aluminium or galvanised thin sheet steel. Common pipe lengths are 6 m and 9 m. Mainlines are usually

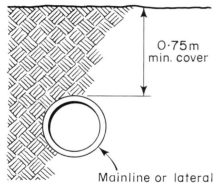

3.13 *Permanent pipes laid below ground level*

0·75 m min. cover

Mainline or lateral

and replacement costs. Remember:

1 Always move pipes with care. Do not throw them as they dent and buckle very easily.
2 Do not walk on the pipes or drive vehicles across them. Use a pipe bridge to cross a pipeline (figure 3.14).
3 Do not leave pipes in contact with harmful or corrosive chemicals, eg fertilisers, pesticides, bird droppings.
4 Use specially built trailers to transport pipes and other equipment (figure 3.15).

larger than laterals with diameters varying from 75 – 200 mm. Laterals which are moved frequently have diameters ranging from 50 – 125 mm.

Care and maintenance

Permanent underground pipes usually require no maintenance once they are installed provided they are adequately protected against corrosion. This is usually done by dipping them in asphalt or wrapping with bitumenous felt, plastic or fibreglass. Steel pipes can also be protected by galvanising.

Portable pipes, particularly aluminium, are easily damaged with continual rough handling. As they form a large part of the cost of any sprinkler system, they should be treated with care to avoid expensive repair

3.15 *Trailer built to carry irrigation equipment*

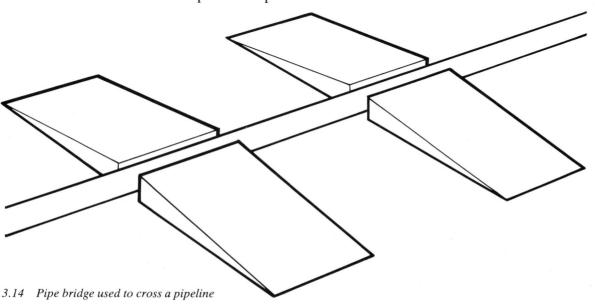

3.14 *Pipe bridge used to cross a pipeline*

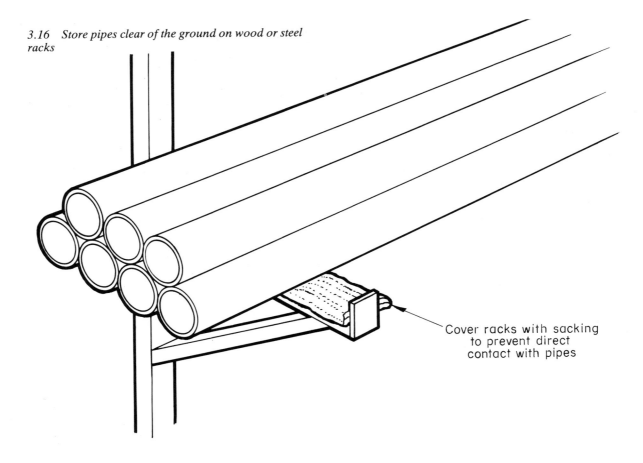

3.16 Store pipes clear of the ground on wood or steel racks

Cover racks with sacking
to prevent direct
contact with pipes

5 At the end of the season, store pipes under cover if possible, clear of the ground on wood or steel racks (figure 3.16). Provide sufficient support to avoid pipes sagging in the middle. Do not treat wooden racks for worm or rot as many of the chemicals used contain copper which damages aluminium.

6 If pipes are stored in the open make one end of the rack slightly higher than the other so that any water collecting in the pipes will drain away.

7 Always repair pipes which have been dented before storing. If damage is not too severe simple hand tools are available to hammer out dents (figure 3.17).

3.6 Pipe couplers

Portable mainlines and laterals are fitted with special joints or couplers so that pipes can be coupled and uncoupled quickly and easily. There are two types:

    Pressure seal couplers
    Mechanical seal couplers

Pressure seal couplers

These are the most common type (figure 3.18). One pipe, fitted with a hook or ring, slides into a socket or coupler welded to the next pipe. By rotating the pipe, the hook locks into a groove on top of the pipe holding the two firmly together. They are uncoupled by lifting the hook clear of the groove and withdrawing the pipe. There is a wide tolerance on the angle at which the pipe can be presented to the coupler. The pipe can thus be held more conveniently midway along its length with no need for the operator to be at the connection (figure 3.19).

Leakage at the coupler is prevented by a V-shaped rubber ring which fits into a groove inside the socket (figure 3.18). When water flows through the pipes, the ring opens and seals the gap between the pipes. When the water is turned off the ring closes and water leaks out of the joints. This helps to drain down the pipes making them easier to uncouple and move.

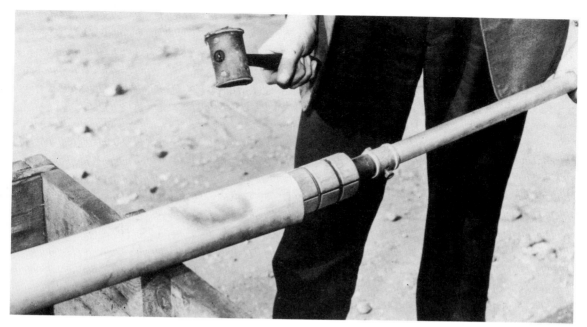

3.17 *Simple hand tools for hammering out dents in aluminium pipes*

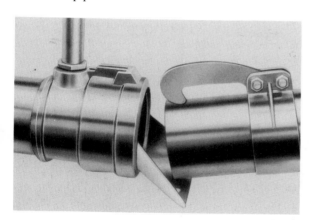

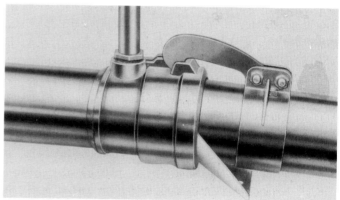

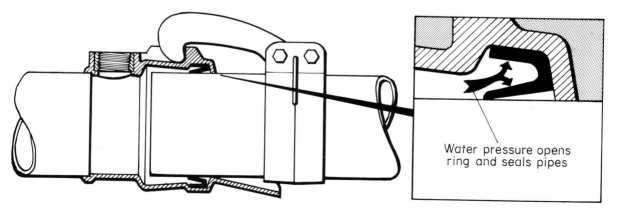

Water pressure opens ring and seals pipes

3.18 *Typical pressure seal coupler*

3.19   *Joining pipes with pressure seal couplers*

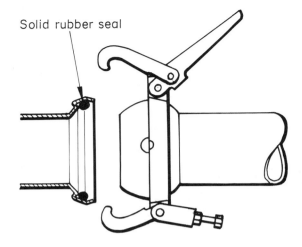

Solid rubber seal

Mechanical seal couplers

These are normally only used on galvanised steel pipes (figure 3.20). A solid rubber sealing ring is squeezed tightly between the two ends of the pipes by the action of a lever arm. This joint prevents leakage at anytime. It must be opened to drain down the pipes so that they can be moved.

Both couplers allow small changes in alignment to be made between pipes (figure 3.21). This gives flexibility to the mainline and laterals to follow the natural ground profile and get around natural obstacles.

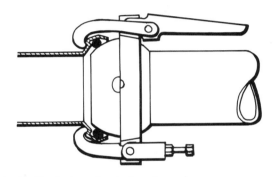

3.20   *Typical mechanical seal coupler*

Care and maintenance

1   Couplers are an important part of the pipe system. Treat them with the same care and attention as the pipes.

2   Sealing rings are made of natural rubber and prevent leakage at pipe joints. Inspect them each season for signs of perishing or damage. Damage occurring during use will be evident from leakage at the coupler.

3   At the end of the season, remove sealing rings from couplers (figure 3.22). To avoid perishing, brush in clean water, allow to dry and store in boxes away from light in a cool place. Pack carefully to avoid distorting rings and sprinkle them with french chalk to absorb any remaining moisture. Alternatively store them in a large jar or tin filled with water.

4   When replacing seals again make sure that the ring seats evenly around the inside of the coupler and that no debris is trapped between the ring and the wall.

3.21   *Couplers allow small changes in pipe alignment to be made between pipes*

3.22 *Removing sealing ring from a pressure seal coupler. (a) Removing seal (b) Pressure sealing ring*

3.7 Valves

Valves are used to control the flow of water and pressure in mainlines and laterals. They are sometimes referred to as hydrants.

On portable systems, valves are used at connections between laterals and the mainline. A special coupler is used which replaces the normal pipe coupler (figure 3.23). An elbow fits over the top of the valve and can be aligned horizontally in any direction. It is held tightly in place by two latches. A socket at the base of the hand-wheel spindle fits over the raised boss on top of the valve. It is opened by turning the hand-wheel anti-clockwise. Notice how the valve closes against the pressure in the mainline. Originally valves were made to close with the pressure. This made them easier to seal but it meant they could be closed rapidly with the help of the pressure and this caused waterhammer problems (see page 25). It is much more difficult to close a valve against the mainline pressure and so the job is done more slowly thus reducing the risk of a pipe burst from waterhammer.

A valve elbow is usually fitted with a pressure gauge. The sprinkler operating pressure is set by turning the valve handle until the pressure gauge registers the correct pressure. The lateral connects into the elbow using an ordinary pipe coupler. On completing an irrigation the elbow is moved with the lateral on to the next valve coupler.

A similar valve assembly is used to connect laterals with permanent underground mainlines (figure 3.24).

Care and maintenance
1 Always open and close valves slowly to avoid damage to the pipe system and pumps from waterhammer.
2 At the end of the season, check valve stems and seating pads for pitting or damage and replace where necessary.
3 Do not seal valves tight when not in use to avoid the rubber sticking to the seat. Open them so that there is a small gap between the valve rubber and seat (figure 3.25), large enough to provide a gap but small enough to prevent rodents from entering.

3.8 Filters

Sprinkler nozzles are very prone to blockage if the water supply is not clean. Dirt, grit and weeds can be drawn into the system by the pump or enter the pipes when they are being moved from one setting to the next. To prevent blockage, filters are placed on the suction side of the pump (see page 105) and at various places in the pipe system. A convenient location for a filter in a pipe is at the head of a lateral between the valve elbow and the first section of pipe. It is usually made from thin sheet brass perforated with fine holes (figure 3.26).

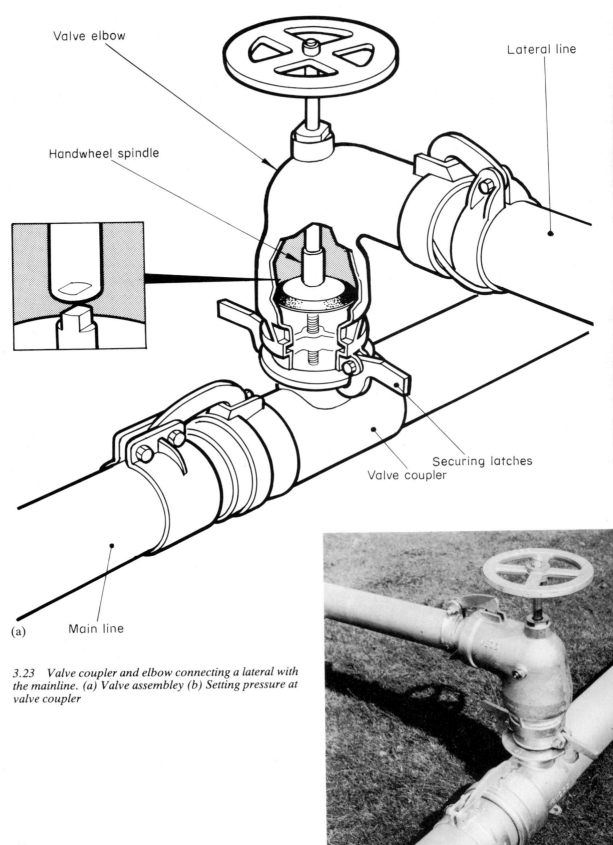

Valve elbow

Lateral line

Handwheel spindle

Securing latches

Valve coupler

(a) Main line

3.23 *Valve coupler and elbow connecting a lateral with the mainline. (a) Valve assembley (b) Setting pressure at valve coupler*

(b)

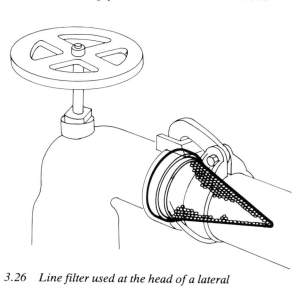

3.25   Leave a small gap between valve rubber and seat

3.24   Valve assembley connecting lateral to an underground mainline

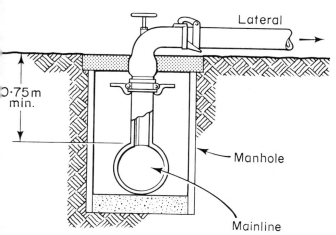

Lateral

0·75m min.

Manhole

Mainline

3.26   Line filter used at the head of a lateral

A common practice with portable laterals is to allow the first flush of water to flow out of the end of the line. This washes out any soil that has entered the pipe during assembly and which may have blocked the sprinklers.

Care and maintenance
1 Filters prevent sprinkler nozzle blockage but can easily become blocked themselves. For portable systems, clean filter every time the pipe is moved. For more permanent systems remove and clean filter each week.
2 If the filter is damaged or deformed, replace it.
3 Blockages can often be detected by checking the operating pressure in the mainline. An undue rise in pressure may be the result of pipe or filters blocking. A drop in pressure just beyond a much higher one should point to the location of the blockage.

3.9 Risers
Risers are small diameter pipes which connect the sprinkler to the lateral (figure 3.27). Pipes from 12-25 mm diameter are used depending on the size of the sprinkler. Connection is made using standard pipe screw threads. The height of the riser is chosen to allow the sprinkler to operate above the crop canopy. Where very tall risers are needed, for example, in maize or sugar cane, stabilizing battens are used (figure 3.28). These hold the riser firmly in an upright position.

Risers are normally connected to the lateral at a pipe coupler. Most couplers are provided with such a connection. When it is not required it can be blanked off with a plug.

In the hop and pipe-grid systems (see pages 30-33) the riser and sprinkler can be disconnected from the lateral while it is still operating. A special riser coupler containing a valve is used which screws into the pipe coupler (figure 3.29). The riser has a plain end which is pushed down into the riser coupler and twisted to lock it in position. This action opens the ball valve and releases water into the riser. When irrigation is complete, the riser is withdrawn and the valve automatically closes again.

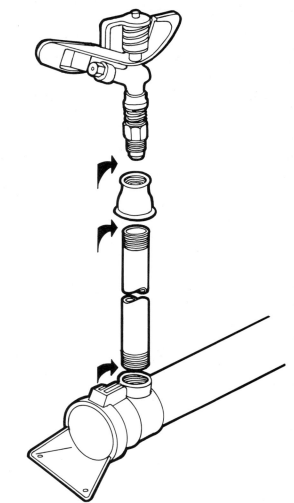

3.27 *Risers connect the sprinkler to the lateral*

Care and maintenance
Risers are usually made from galvanised steel or thick walled aluminium tube for strength. Always use recommended jointing compounds when joining as this makes them easier to remove at a later date. Compound is essential when joining galvanised steel to aluminium to prevent corrosion between the two metals.

3.10 Rotary sprinklers
Rotary sprinklers are of robust construction to withstand high operating pressures and continuous handling in the field (figure 3.30). They are usually made of corrosion resistant bronze alloys or stainless steel although some new sprinklers are being made from strong plastic.

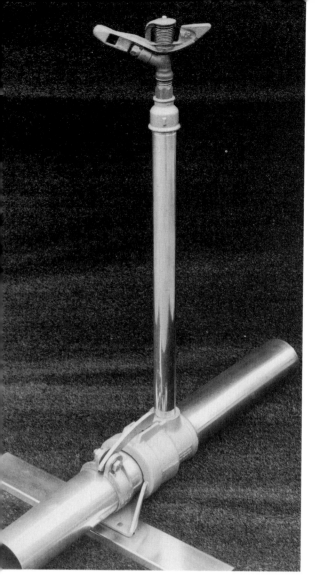

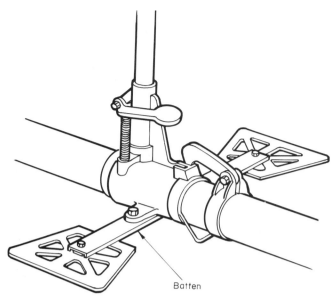

3.28 *Stabilizing battens used on tall risers*

Rotary sprinklers have either one or two nozzles (figure 3.31). A single nozzle sprinkler provides good uniformity and can apply water at low rates. A two-nozzle sprinkler has a main or range nozzle, similar to the single nozzle, and a spreader nozzle which irrigates close to the sprinkler. This improves the uniformity of application but watering rates are generally higher and the spray is more affected by wind.

Range nozzles normally vary from 3-10 mm in diameter. Spreader nozzles are smaller in diameter than the range nozzle. Some sprinklers have small removable plastic inserts behind the range nozzle. These contain several vanes which 'straighten' the flow in the sprinkler and suppress turbulence. A similar device is described in more detail on page 60.

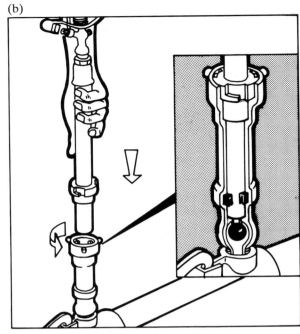

3.29   Special riser coupler allows riser to be connected to lateral while operating (a) Riser coupler

(b) Riser is pushed down and twisted to lock thus releasing water

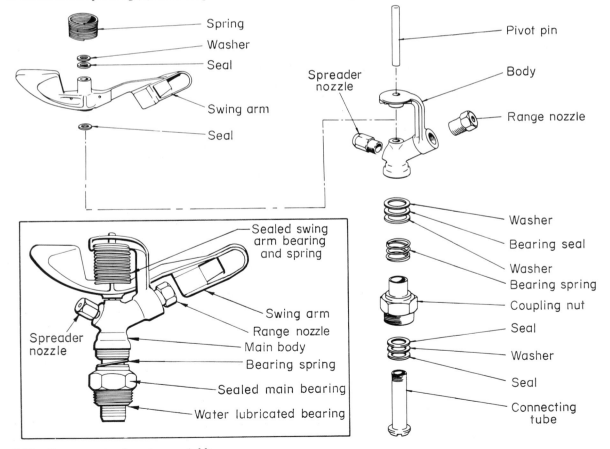

Spring
Washer
Seal
Swing arm
Seal

Pivot pin
Spreader nozzle
Body
Range nozzle

Sealed swing arm bearing and spring
Swing arm
Spreader nozzle
Range nozzle
Main body
Bearing spring
Sealed main bearing
Water lubricated bearing

Washer
Bearing seal
Washer
Bearing spring
Coupling nut
Seal
Washer
Seal
Connecting tube

3.30   Components of a rotary sprinkler

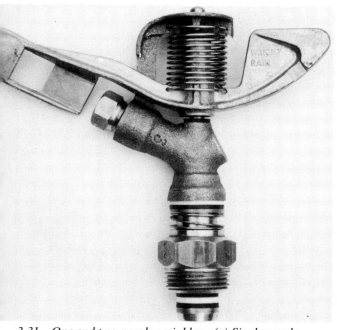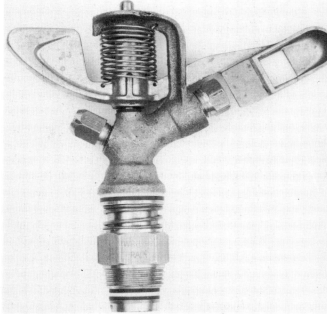

3.31 One and two-nozzle sprinklers. (a) Single nozzle   *(b) Two-nozzle*

Care and maintenance

Rotary sprinklers are precision made to close tolerances in order to achieve good uniform irrigation and correct drop size distribution. To ensure that they continue to operate satisfactorily, frequent checks should be made on their performance.

If sprinklers are moved regularly the following checks should be made daily:

1 Ensure that nozzles are completely free of any obstruction. **Do not** clear obstructions with a sharp metal object such as a screw driver as this can damage the bore of the nozzle.
2 When the sprinklers are fitted with removable flow straightners, take care when removing nozzles to prevent their loss.
3 Ensure that the swing arms are free to rotate.
4 Check that the main body of each sprinkler is free to rotate on its main bearing.

Sprinkler main bearings are self-lubricated by the irrigation water. It is important that no mineral-based oil is applied to them as it will perish the rubber seals. If the water appears to be leaking from the bearing it may be the result of grit lodged between the bearing surfaces. This can be cleared by pushing down the sprinkler on to the coupling nut so that the water pressure will dislodge the grit.

At the end of the irrigation season the following checks should be made before laying up the equipment in storage:

1 Check all sprinklers for wear in the main and swing arm bearings. Wash and brush them in clean water to remove any grit.
2 Check nozzle bore for damage and wear. Silt and sand particles in irrigation water can cause wear and increase the size of the bore. This will affect the performance of the sprinkler. Nozzle wear can easily be checked using the shank of a twist drill, the same size as the nozzle (figure 3.32).
3 Check tension spring on swing arm. If a new one is fitted, set the tension to give the recommended speed of rotation using a spring gauge (figure 3.33).
4 General sprinkler performance can be checked using a simple test rig made from a large oil drum (figure 3.34). Remember to observe performance only at the recommended operating pressure.

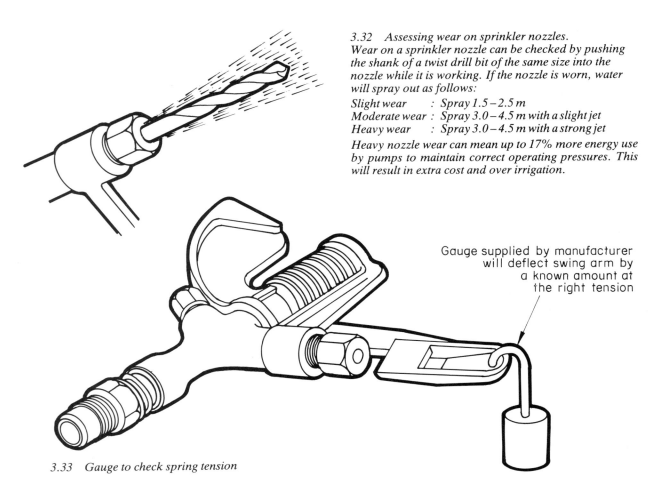

3.32 *Assessing wear on sprinkler nozzles.*
Wear on a sprinkler nozzle can be checked by pushing the shank of a twist drill bit of the same size into the nozzle while it is working. If the nozzle is worn, water will spray out as follows:

Slight wear   : Spray 1.5 – 2.5 m
Moderate wear : Spray 3.0 – 4.5 m with a slight jet
Heavy wear   : Spray 3.0 – 4.5 m with a strong jet

Heavy nozzle wear can mean up to 17% more energy use by pumps to maintain correct operating pressures. This will result in extra cost and over irrigation.

Gauge supplied by manufacturer will deflect swing arm by a known amount at the right tension

3.33 *Gauge to check spring tension*

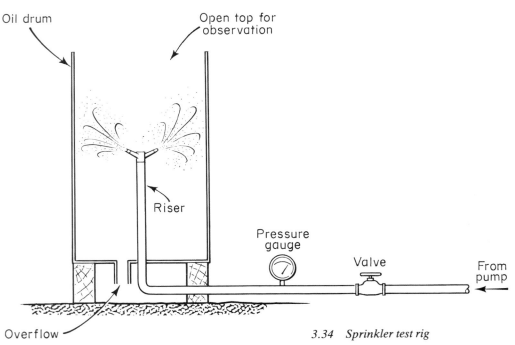

Oil drum

Open top for observation

Riser

Pressure gauge

Valve

From pump

Overflow

3.34 *Sprinkler test rig*

3.11 Pressure regulators

The satisfactory performance of rotary sprinklers depends to a large extent on operating pressure. In some systems it may be difficult to maintain the right pressure at every sprinkler. An example of this might be in an area where the ground surface is very uneven (figure 3.35). In such cases a pressure regulator can be used to control the pressure at each sprinkler (figure 3.36). This device fits into the top of the riser just under the sprinkler. Normally pressures P_1 and P_2 are equal and close to the sprinkler operating pressure. If the pressure P_1 rises then P_2 also starts to rise. This pushes down the sleeve closing up the waterway and reducing the flow to the sprinkler. The effect of this is to reduce the pressure P_2 to the sprinkler operating pressure which is less than P_1. If P_1 starts to fall then P_2 also falls. The sleeve opens and allows more flow through to maintain the pressure P_2.

The regulated pressure P_2 is controlled by a spring in compression inside the sleeve. This can be adjusted using a small screw. For example, if a higher pressure P_2 is required then the screw is turned clockwise increasing the spring compression. This stops the movement of the sleeve until the higher pressure P_2 is reached.

Pressure regulators will only reduce the pressure in the lateral to the correct sprinkler operating pressure. They will not increase it when the lateral pressure is inadequate.

Care and maintenance

1 Pressure regulators are usually pre-set by the manufacturer to suit a particular sprinkler system. They should not require adjustment in the field.
2 Always make sure that the regulator is placed the right way up in the riser tube.

Larger pressure regulators are also available to control pressures in laterals and mainlines. They work in the same way as the small sprinkler regulators.

3.12 Pipe specials

Specials are short lengths of pipe or fittings used in mainlines and laterals (figure 3.36).

Pipe bends or elbows

Elbows or bends are used when a pipe needs to change direction. Typicals bends are for 90° and 45°. They are usually reversible so that maximum use can be made of them.

Reducers

Reducers are short lengths of pipe used to connect pipes of different diameter.

Tees

A 'Tee' enables a branch main to be taken of a mainline, usually at 90° to the mainline.

End plugs

End plugs are fitted at the ends of mainlines and laterals. Figure 3.37 shows how these and other components are used in a typical portable sprinkler system.

FIELD MANAGEMENT

3.13 Moving pipes and sprinklers

One of the most frequent tasks undertaken with conventional irrigation systems is to move sprinklers and laterals in the field. Portable systems, for example, may involve as many as four moves a day for each lateral during a busy irrigation season. It is useful therefore to observe a number of basic rules to see that the job is carried out quickly and with the minimum risk of damage to both the crop and the equipment.

1 When the lateral is laid in position it is good practice to flush the pipe out before fitting the end plug. This will wash out any soil that has found its way into the pipe during assembly.
2 When irrigation begins the pipes are full of air. This has to escape before the pipe can fill with water and may take several minutes. Air usually escapes through the sprinklers and pipe joints or a valve can be provided for this purpose at the far end of the lateral.
3 Open the valve coupler controlling flow into the lateral very slowly, to avoid placing undue strain on the pumps and water hammer at the far end of the lateral.

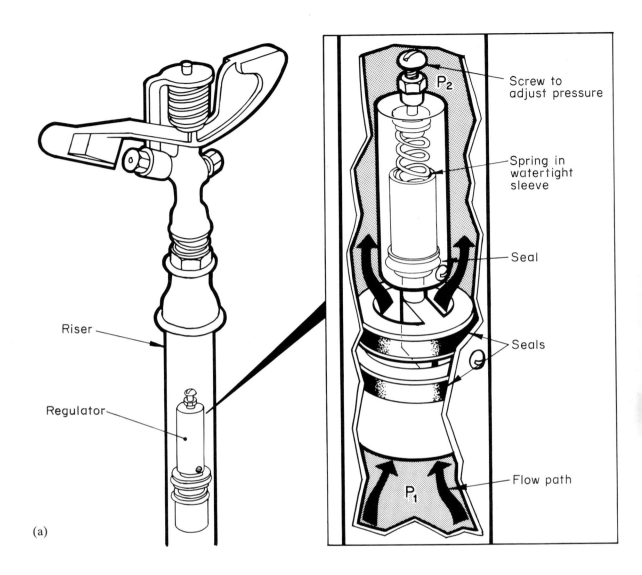

Screw to adjust pressure

Spring in watertight sleeve

Seal

Seals

P_2

P_1

Flow path

Riser

Regulator

(a)

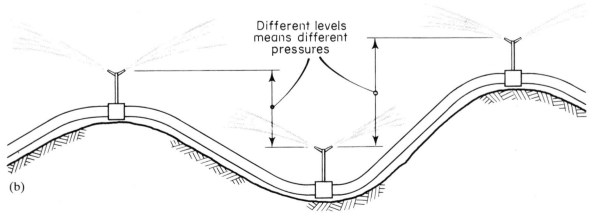

Different levels means different pressures

(b)

3.35   *Controlling sprinkler operating pressure (a)*
Pressure regulator (b) Pressure changes across uneven
ground

a)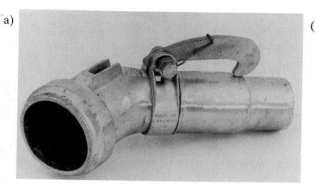

(a)

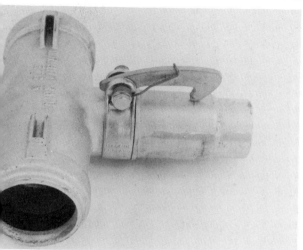

3.36 *Pipe specials (a) 90° and 45° bends (b) Tee* (d)
(c) Reducer (d) End plug

4 Set the sprinkler operating pressure by turning the valve until the pressure gauge shows the correct pressure (figure 3.23).

5 On completing the irrigation close the valve coupler on the mainline slowly to avoid water hammer problems and allow the lateral to drain down.

6 Disconnect the valve elbow and the sections of pipe and move to the next irrigation position. The most efficient way to do this is to follow the paths marked in figure 3.37.

7 Re-assemble the pipe, remembering to clean and insert the line filter.

8 **Always** carry pipes horizontally (figure 3.39). Never up-end them. There may be overhead power cables about and a danger of electrocution.

A lateral moving technique sometimes used by experienced irrigators is to open the valve coupler a little when the first two sections of pipe are in place. This allows a small stream of water to flush out any soil in the pipes. As each section of pipe is put into place the flushing action continues. The last section of pipe and the end plug can also be connected before the pressure builds up in the system. The irrigator then walks back along the lateral checking for sprinkler blockages, leaky seals, and straightening risers which may have fallen over at an angle. On reaching the mainline he then opens up the valve to obtain the full working pressure. A quick check with a pressure gauge on the first sprinkler confirms this.

49

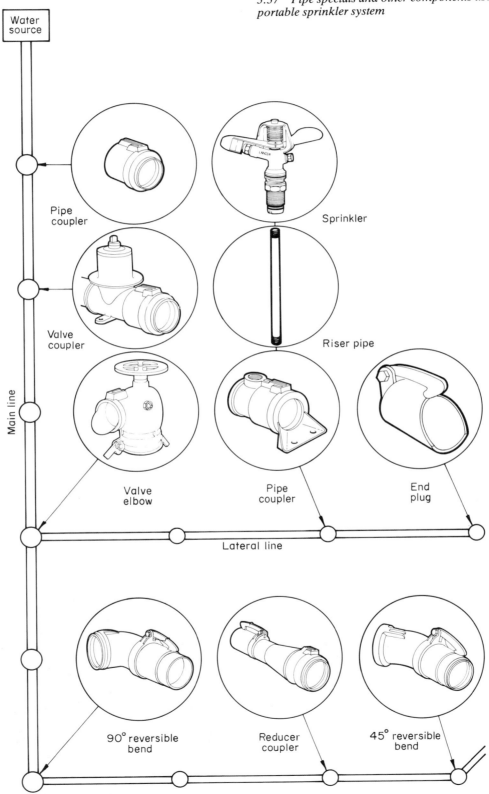

Water
source

Pipe
coupler

Sprinkler

Valve
coupler

Riser pipe

Main line

Valve
elbow

Pipe
coupler

End
plug

Lateral line

90° reversible
bend

Reducer
coupler

45° reversible
bend

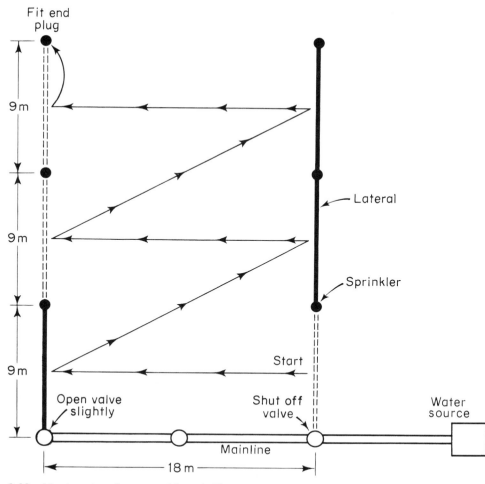

Fit end plug

9 m

Lateral

9 m

Sprinkler

9 m

Start

Open valve slightly

Shut off valve

Water source

Mainline

18 m

3.38   Moving pipes for a portable sprinkler system

3.39   Moving portable pipes

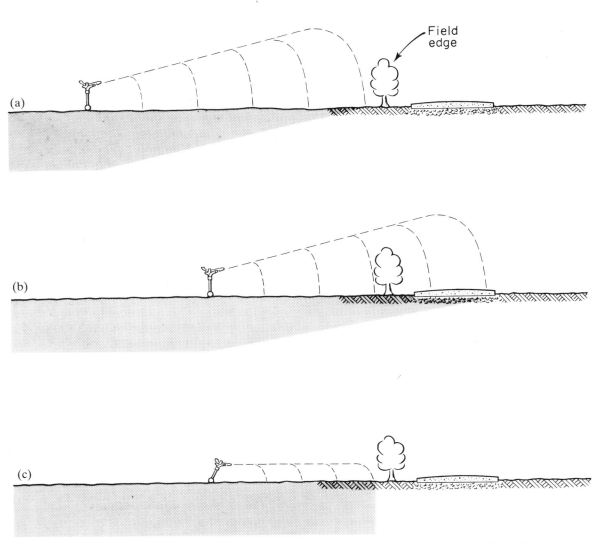

(a)

(b)

(c)

3.40 Setting sprinklers near field boundary
(a) Poor distribution near field edge
(b) Improved distribution but water spills onto road
(c) Good distribution and no spillage

3.14 Field boundaries

In view of the triangular shape of the sprinkler distribution pattern there is always a problem near the edge of fields. If the lateral is placed away from the edge it is under irrigated. If it is placed near to the edge then water spills over into the road or the next field. (figure 3.40). One solution is to place the lateral near the edge of the field and tilt the riser slightly so that it throws the water towards the ground. This avoids irrigating outside the field and improves uniformly near the boundary.

4

MOBILE RAINGUN SYSTEMS

4.1 Introduction

Mobile raingun systems use a large rotary sprinkler operating at high pressure to irrigate large areas. The term *raingun* is used to describe them because of the large size of sprinkler used and its ability to throw large quantities of water over wide areas. Although simple hand-moved equipment similar to that used in conventional sprinkler systems is available for rainguns, it is more common for them to be mounted on a portable carriage which moves continuously across the field while irrigating. Machines like this are often called *travellers*. They have become very popular in recent years because of their relatively low capital cost per hectare and low labour requirements.

Rainguns normally operate at high pressure from 5-10 bar (75-145 lbf/in.^{2}) with discharges ranging from 40-120 m^3/h. They can irrigate areas up to 100 m wide and 400 m long (4 ha) at one setting. Application rates vary from 5-35 mm/h.

There are two main types of system:
Hose-pull system
Hose-reel system

4.2 Hose-pull system

The hose-pull machine has a raingun mounted on a wheeled carriage (figure 4.1). Water is supplied through a flexible hose up to 200 m long and 50-100 mm diameter which is pulled along behind the machine. In a typical layout for a hose-pull system (figure 4.2) the mainline is laid across the centre of the field from the pumping station. A strip up to 400 m long can be irrigated at one setting although the flexible hose may only be 200 m long.

The raingun carriage is positioned at the start of its first run (figure 4.3). The flexible hose is laid along the travel lane and connected to the raingun and the valve coupler on the mainline. Care is needed at this stage to avoid kinks and twists in the pipe as they will obstruct the flow of water. A loop is formed in the hose behind the machine.

A steel guide cable on the sprinkler carriage is pulled out to the far end of the field and firmly anchored. The valve coupler is slowly opened to start the irrigation. The raingun carriage is moved either by a 'water motor' powered from the water supply using a piston or turbine drive (see page 61), or an internal combustion engine. This slowly turns a winch which winds in the guide cable and pulls the raingun across the field.

The application rate is controlled by the pressure at the raingun. The forward speed of the machine controls the depth of water applied. Typical machine speeds vary from 10-50 m/h. The faster the machine travels the smaller the depth of water applied.

Once the machine is operating it should not require any supervision for many hours. At the end of a run it stops automatically. Some hose-pull machines also have a mechanism which shuts down the main water supply to the raingun. On simpler machines an operator needs to be on hand to stop the pump. Labour is required only to reposition the hose and machine and to start the next run. This usually takes one man with a tractor approximately one hour.

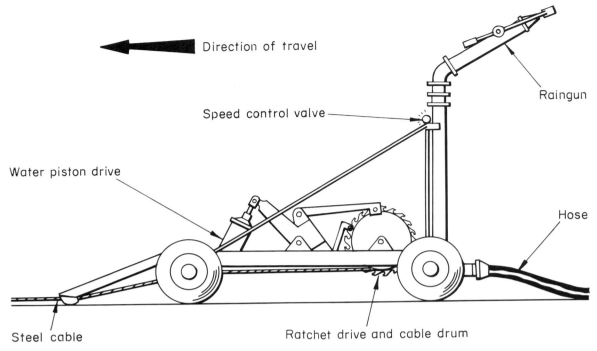

Direction of travel

Speed control valve

Water piston drive

Raingun

Hose

Steel cable

Ratchet drive and cable drum

4.1   Hose-pull machine

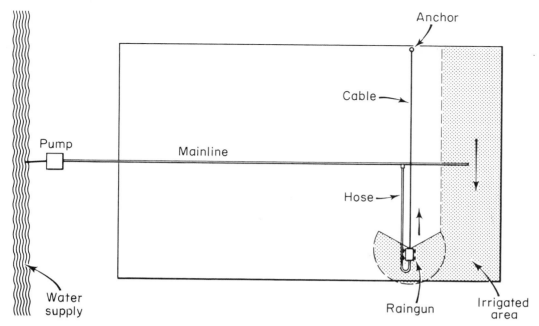

4.2   Typical field layout for a hose-pull system

4.3   Setting up hose-pull system (a) Positioning raingun carriage (b) Laying out the flexible hose (c) Cable anchor firmly fixed at far end of run   (b)

(a)

(c)

4.3 Hose-reel system

The hose-reel machine has a raingun mounted on a sledge or wheeled carriage (figure 4.4). Water is supplied through a more rigid hose than that used for the hose-pull although it is still flexible enough to be wound on to a large reel. The hose is used to pull the raingun towards the hose reel positioned at the edge of the field. Machines are available with hose lengths ranging from 200-400 m.

In a typical layout for a hose-reel system (figure 4.5) the mainline is laid out across the centre of the field from the pumping station. The hose-reel is placed close to the mainline at the start of the first run and connected to the water supply. The raingun is slowly pulled out across the field by tractor and the hose allowed to uncoil from the reel. Only the hose length needed is pulled out, the surplus remains neatly coiled on the reel. The pump is started and the valve coupler slowly opened to start the irrigation. The raingun is slowly pulled back across the field by winding the hose on to the hose-reel. Power to drive the hose-reel can be provided by a water motor, an internal combustion engine or the power take-off point on a tractor. At the end of a run the hose-reel automatically stops winding. On some machines a mechanism also shuts down the main water supply to the raingun.

When the hose-reel is used in the center of the field it is turned through 180° and the raingun pulled out to start the next irrigation run, a job carried out simply by one man and a tractor. When irrigation is completed in this position the hose reel and raingun are towed by tractor to the next field location. For small fields the mainline may be placed along one edge, provided the hose is long enough. Application rates and machine speeds are similar to the hose-pull system.

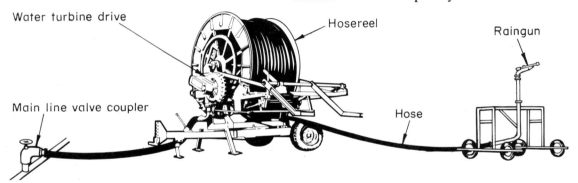

Water turbine drive   Hosereel   Raingun

Main line valve coupler   Hose

4.4   Hose-reel machine

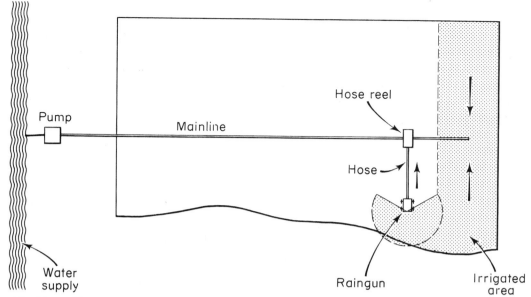

4.5   *Typical layout for a hose-reel system*

4.4 Rainguns

Rainguns are large rotary sprinklers similar in many ways to the small ones described page 00. They are usually mounted on sledges or wheeled carriages which can be adapted to suit the various furrow or row spacings and crop heights.

They are of robust construction to withstand the large forces produced by the high discharge rates and operating pressures. There are two types:

    Swing-arm raingun
    Water-turbine raingun

Swing-arm raingun

This operates in a similar manner to the small rotary sprinkler (figure 4.6). It rotates by means of a drive spoon on the end of a swing-arm which is free to move up and down. The spoon is shaped so that the water jet pushes the swing-arm downwards out of the flow. At the same time it is pushed sideways causing the raingun to turn slightly. Once clear of the flow the swing arm is so balanced that it returns to interrupt the jet again. The raingun then receives another sideways impulse. This action is repeated in a steady beating motion causing the raingun to slowly rotate. The spoon also helps to break up the water jet into fine droplets.

The speed at which the raingun rotates is controlled by the angle of the drive spoon and an adjustable friction brake. One complete revolution can take from 2 to 5 minutes.

Rainguns can irrigate through a full circle but sector rainguns which irrigate only part of a circle behind the machine are most common. This ensures that the machine always moves on a dry travel lane (figure 4.7). When a sector raingun reaches the end of its circular path it reverses rapidly again to the beginning. This is achieved by a series of smooth cams known as sector stops. When nearing the end of its path the cam roller rides on to the sector stop and engages the reversing arm with the water jet. The force of the water on the arm causes the raingun to return rapidly. On returning, the same cam roller rides on to a second sector stop which disengages the reversing arm from the jet. The raingun is then set to start irrigating again. The positions of the sector stops are adjustable so that any size of circular arc can be irrigated. A typical sector covers up to 270°.

A rapid return of the raingun at the end of a sector avoids unnecessary wastage of water but can create problems by over-riding the disengaging stop. In order to produce a sprinkler pattern which is symmetrical the

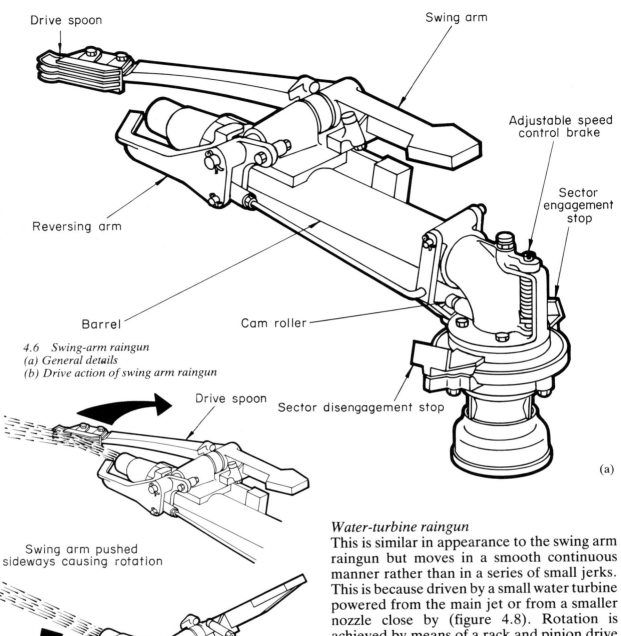

Drive spoon

Swing arm

Adjustable speed
control brake

Sector
engagement
stop

Reversing arm

Barrel

Cam roller

4.6 Swing-arm raingun
(a) General details
(b) Drive action of swing arm raingun

Drive spoon

Sector disengagement stop

(a)

Swing arm pushed
sideways causing rotation

Swing arm deflects
(b) downwards out of flow

Water-turbine raingun

This is similar in appearance to the swing arm raingun but moves in a smooth continuous manner rather than in a series of small jerks. This is because driven by a small water turbine powered from the main jet or from a smaller nozzle close by (figure 4.8). Rotation is achieved by means of a rack and pinion drive connected to the turbine through a small gear box. The speed at which the raingun rotates is controlled by the speed of the turbine wheel.

Water turbine rainguns can also be used for sectoring and stops are provided to limit movement. When the raingun reaches the end of its circular path it runs up against a sector stop. This causes the whole body of the turbine and gear box to rotate about the axis of the drive shaft so that the other side of the turbine wheel lines up with the water jet. This drives the turbine in the opposite direction

disengaging stop is sometimes positioned to allow for the inertia for the raingun. However, the speed of the gun reversing action is adjustable and should be checked and reset if seen to be very forceful.

and so reverses the movement of the raingun. The raingun slowly moves back along its path until it reaches the other sector stop. This again reverses the turbine and the movement of the raingun. The raingun thus moves slowly back and forth while irrigating; unlike the swing-arm raingun which moves slowly in one direction and reverses rapidly in the other.

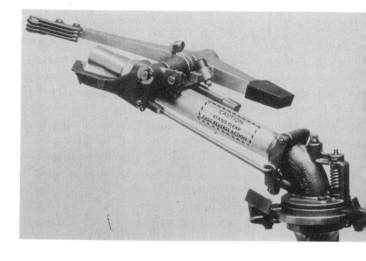

General features

Rainguns are fitted with either taper-bore or ring nozzles (figure 4.9(a)). Taper nozzles normally produce a good water jet which will penetrate the air more easily and is less affected by wind. They also have a much greater throw than ring nozzles. Ring nozzles however, provide better stream break up at lower operating pressures – an important

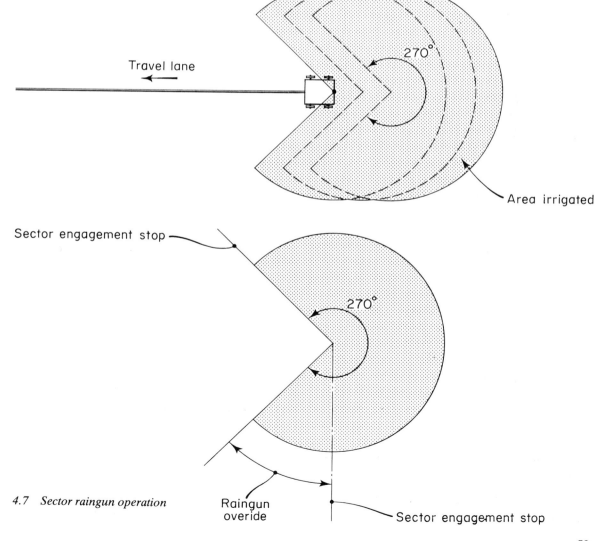

4.7   Sector raingun operation

factor on delicate crops. They are cheaper and give greater flexibility in size selection than taper nozzles. This can be particularly useful when a system is first installed and the operator is unsure of the best size to use or where a raingun is used on several different crops requiring frequent changes of nozzle. Typical nozzle sizes vary from 15-50 mm diameter.

The trajectory angle for rainguns varies between 15° and 27° (figure 4.9(b)). Generally, the higher the angle the further the water will travel for a given operating pressure. If adequate operating pressure is available, low angles are preferred as this reduces the height through which water drops have to fall and helps to reduce the effects of wind distortion on the sprinkler wetting patterns.

As rainguns operate at high pressures it is important that the jet of water leaves the nozzle relatively undisturbed. Any turbulence in the flow will reduce the throw of the sprinkler. Turbulence occurs through poorly designed pipework, abrupt changes in pipe size and roughness on the inside of the pipes. Many rainguns now have vanes in the main barrel leading to the nozzle which 'straighten' the flow and suppress turbulence (figure 4.9(c)).

Safety
Rainguns operating at high pressure can be dangerous. Treat them with great care.
1 **Always** read the manufacturers' instructions before operating or making adjustments to the raingun.
2 Never make adjustments or carry out servicing while the raingun is in use. Make one adjustment at a time and in small increments.
3 Keep well clear of the raingun in operation. The swing arm type reverse action is fast

4.8 Water-turbine raingun

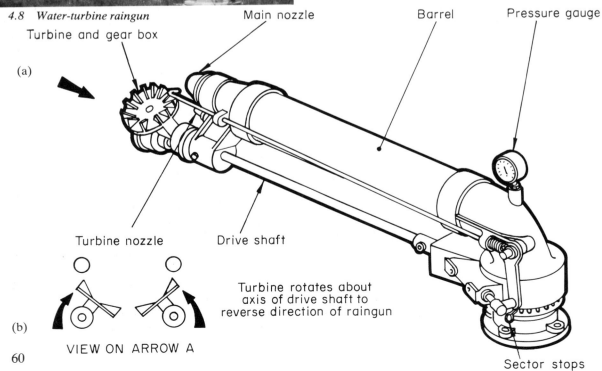

60

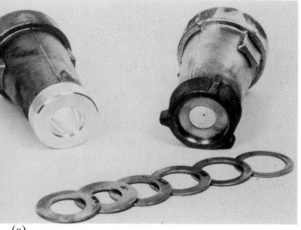

(a)

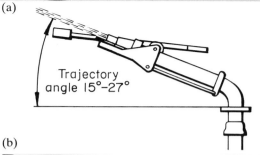

Trajectory angle 15°–27°

(b)

(c)

4.9   General features of rainguns (a) Taper and ring nozzles (b) Trajectory angle (c) Vanes in barrel

and powerful and can cause serious injury (see notice on barrel of raingun.).

4 Stand well clear of the high velocity water jet.

5 Never direct jet onto roadways or power lines (see page 75).

6 Check to see that all nuts, bolts and pipe fittings are tightened securely before starting.

Care and maintenance

The moving parts of most modern rainguns are fitted with sealed lubricated bearings which are designed to last the life time of the equipment and usually require no maintenance. However, weekly applications of grease on the rollers and sector stop faces will ensure trouble-free operation.

Trouble-shooting

Typical problems that occur with rainguns are:

1 Stream from the nozzle is turbulent and breaks up too easily. This is usually caused by debris trapped in nozzle or barrel.

2 Rotation speed is too fast or too slow. The speed brake or drive mechanism needs adjusting.

3 Drive spoon on swing-arm raingun does not swing properly. The nozzle or barrel may be blocked with debris or there is insufficient operating pressure or the drive spoon bearing is damaged.

At the end of the season, clean and inspect the raingun in accordance with the manufacturer's instructions. Make any necessary adjustments and replace worn parts as required.

4.5 Drive mechanisms

Although mobile rainguns can be powered by an internal combustion engine or in some cases a tractor power offtake, it is common practice to use a 'water motor' powered from the main water supply. There are two types:

Hydrostatic drive

Hydrodynamic drive

Hydrostatic drive

This drive depends upon the pressure in the water supply to push a piston or a set of bellows which is connected to a ratchet device. The piston type uses a double acting piston (figure 4.10). A small amount of water is taken from the mainline and is supplied alternately to either side of the piston causing it to rise and fall. The movement rocks the drive arm allowing the two ratchet linkages to pull alternately on the toothed ratchet wheel. On the upstroke of the piston the lower linkage turns the wheel, while the upper

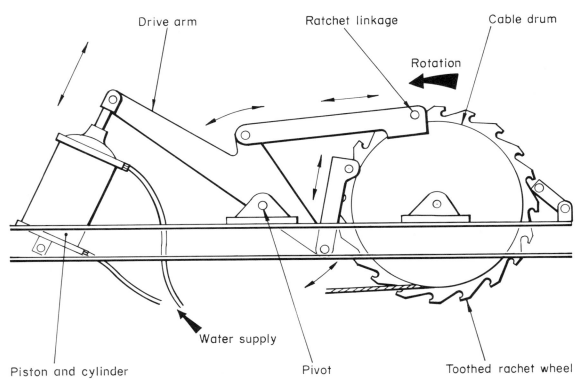

Drive arm   Ratchet linkage   Cable drum

Rotation

Piston and cylinder   Water supply   Pivot   Toothed rachet wheel

4.10   Water piston and ratchet drive on a hose-pull raingun

linkage slides forward to the next tooth. On the downstroke the upper linkage turns the wheel while the lower one slides up to the next tooth. A retention pawl is used to prevent the toothed wheel from rotating backwards.

Water is supplied through a valve and a fine filter is used to stop any dirt or grit entering the system and damaging the control valves (figure 4.11). The main control valve is a linear slide valve which diverts the supply to either side of the piston and also allows the non-pressure side of the cylinder to empty through the water outlet. The valve is operated by sliding the spool inside the valve body so that the appropriate pipe couplings are linked together. The spool can be moved mechanically or hydraulically.

This type of drive is normally used on a hose-pull machine (figure 4.1). The toothed wheel is connected directed to a winch drum which winds in the steel cable and pulls the machine along.

The travel speed of the machine is controlled by adjusting the water flow through the system. A valve on the water outlet pipe is provided for this purpose. As the valve is opened, more flow passes through the drive and the machine moves faster. The water is used to drive the piston is sprayed through the water outlet just below the raingun itself, and irrigates land close to the machine (figure 4.1).

The machine is stopped at the end of a run by automatically closing the water outlet valve. A clip, bolted to the steel cable is made to engage on a trigger which closes the valve (figure 4.12). This mechanism, of course, only stops the machine moving forward and does not shut down the main water supply to the raingun. This would normally be done manually. Machines are available with valves that shut down the whole supply automatically. In such cases it would be necessary to include a by-pass valve on the pump to divert water away from the machine or a pressure sensitive switch which would shut down the pump. An example of this type of system is shown on the water turbine drive.

A simiiar drive system has also been developed using flexible bellows instead of a piston. (Figure 4.13) The bellows expand and contract as water is fed through it and this action rocks the drive arm in the same manner as the piston drive.

Hydrodynamic drive
This drive depends upon the discharge from the mainline to drive a small turbine wheel connected to the hose reel through a gear box or pulley system. There are two ways of driving the turbine:
    Partial flow
    Full flow
When only part of the flow is used, water is diverted from the mainline in a similar manner to the hydrostatic drive. (figure 4.14). After going through the turbine the flow is passed back into the mainline. This avoids the problem of spilling this water close to the machine. The speed of the turbine and hence the speed of travel of the raingun is controlled by the amount of water passing through the turbine. There are disadvantages with this system. As only a small amount of flow is used the pipes tend to be small and are easily blocked. Careful filtration is essential to avoid this. The turbine also lacks power to overcome sudden changes in load on the winding hose reel and this will affect the travel speed of the raingun. Such changes often occur when the raingun is being pulled uphill or across rough ground.

When the whole flow in the mainline is used, a much larger turbine and pipe sizes are used. This reduces the problem of blockage and eliminates the need for filtration. Because the flow through the turbine is now much greater it responds better to changes in loading on the winding hose reel. The speed of the turbine is controlled by a variator or variable speed pulley similar to the kind used on a combine-harvester.

Figure 4.15 shows how this type of drive is used on a hose-reel machine. The turbine is linked to the reel through a variable speed pulley and a gearbox. A clutch is used so that the drive can be disengaged manually at any time.

At the end of an irrigation run there are three ways in which the machine can be stopped:

    Divert the flow away from the turbine

    Slowly shut down the main water supply

    Declutch the turbine drive

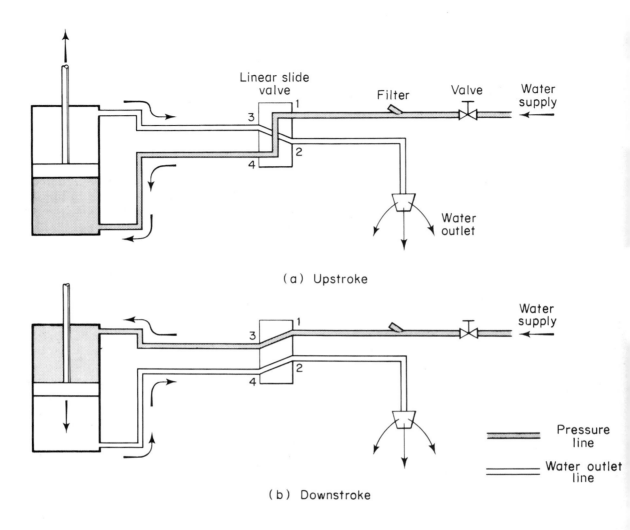

(a) Upstroke

(b) Downstroke

Pressure line

Water outlet line

4.11 *Hydraulic system to control water piston drive.*
(a) Upstroke (b) Downstroke (c) Linear slide valve
feeding water piston

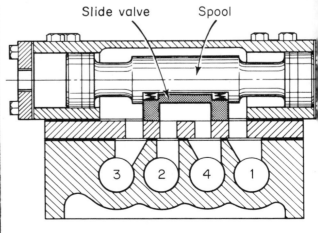

(c) Linear slide valve feeding water piston

4.12 A clip on steel cable used to close water outlet valve

4.13 Bellows drive on a hose-reel raingun

Many machines using one of the first two methods also incorporate mechanical declutching as a safety measure in case the first method fails to operate.

A stopping system which diverts the flow away from the turbine and also declutches the drive is shown in figure 4.16. A fin, fastened to the hose close to the raingun sledge, is used to open a valve on the main line just upstream of the turbine. This diverts all the flow away

from the turbine and raingun and discharges it to waste. The sudden loss of pressure at the raingun can be used to activate a pressure switch which shuts down the pump. The fin is also used to actuate the clutch.

A system which slowly shuts down the main supply is shown in figure 4.15.

4.6 Water supply hoses

Water is supplied to a raingun through a strong flexible hose 50-110 mm in diameter. It is designed to withstand high water pressures and tensile forces which occur as it is pulled across fields.

Hose-pull machine

This machine uses a hose made from strong canvas coated with plastic which protects the pipe from abrasion outside and provides a smooth 'hydraulic' surface inside.

This type of hose lies flat when not in use and is often referred to as a 'lay-flat' hose. It is easily coiled and stored on a reel. On some machines the reel is an integral part of the sprinkler carriage; on others a separate reel mounted either vertically or horizontally on a wheeled frame (figure 4.17) is provided. Both types are usually operated from the power take-off point on a tractor.

The hose is laid out by towing the hose reel across the field by tractor and allowing it to unwind. A loop 2-3 m wide is formed behind the machine. As the raingun moves forward the loop is pulled along behind (figure 4.18). A grassed track known as a travel lane is provided for this purpose as the loop could cause serious damage if pulled through a crop. On some crops it is possible to use a loop lifter. This is a wheeled frame which lifts and supports the loop clear of the crop (figure 4.19) and moves slowly forward with the raingun. The wheel spacing is adjustable to suit various crop row widths.

At the end of a run the hose has to be reeled in before it can be moved to the next position. Before reeling, all water must be removed from the hose. When it is lying on a slope it will drain naturally once it is disconnected from the mainline and raingun. If not, the water must be forced out. On small machines, using hoses less than 75 mm, a squeezing action is used by pulling the hose between

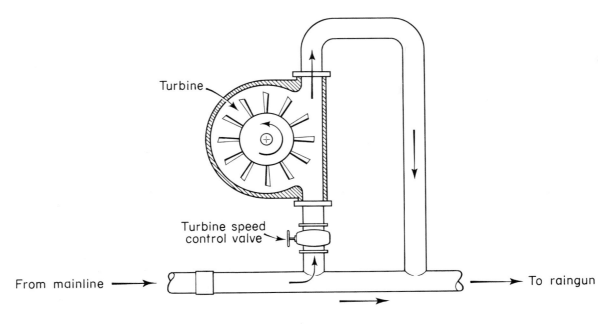

Turbine

Turbine speed
control valve

From mainline →

To raingun

(a)

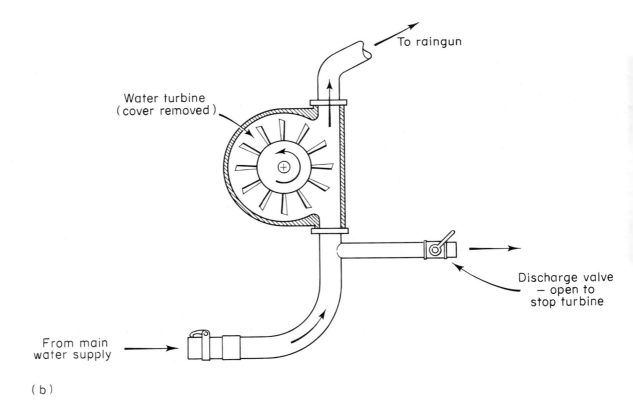

To raingun

Water turbine
(cover removed)

From main
water supply →

Discharge valve
– open to
stop turbine

(b)

4.14   Turbine drives (a) Turbine drive using part of
mainflow (b) Turbine drive using whole flow

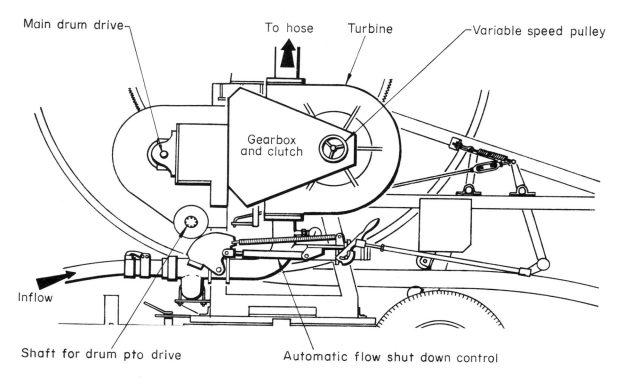

Main drum drive

To hose

Turbine

Variable speed pulley

Gearbox and clutch

Inflow

Shaft for drum pto drive

Automatic flow shut down control

4.15   Turbine drive on hose-reel machine

rollers. For larger hoses a small compressor is used to drive out the water because of the high forces that would be inflicted on the hose if it was pulled full of water.

The maximum length of hose is limited by the pipe diameter and the water pressure available to overcome pipe friction. The pulling force required to move the raingun does not depend on the length of the outlying hose. Only a limited length of hose and the loop are pulled along behind the raingun,

resulting in a relatively constant pull of approximately 1 tonne.

Special couplers are used to connect the hose to the pump and the raingun (figure 4.20). Three cuts are made in the end of the hose so that it can be pushed onto the tapered connector. It is held in place initially by a steel band and then trapped between the inner taper and an outer tapered cover. Pulling the hose tightens the coupling.

Hose-reel machine
This machine uses a more rigid plastic pipe than the hose-pull machine although it is still sufficiently flexible to be wound onto a large reel (figure 4.4).

The hose is laid out across the field by towing the raingun sledge with a tractor and allowing the hose to unwind from the hose reel. This must be done slowly and smoothly to avoid excessive loads on the hose. Care is needed to avoid towing the raingun sledge further than the hose length will allow. When irrigating, the hose is slowly wound onto the reel, pulling the raingun across the field. To ensure a constant speed of travel, the speed at which the hose-reel winds in must gradually slow down. This is because the amount of hose

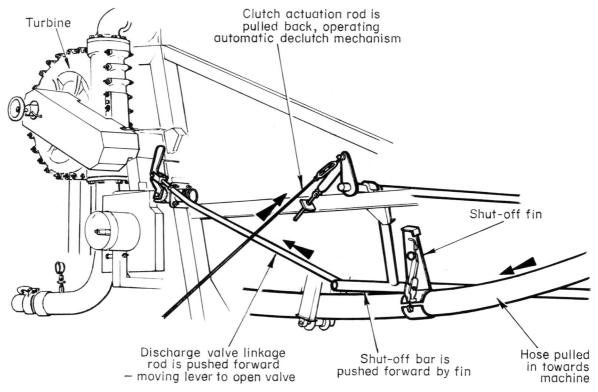

Turbine

Clutch actuation rod is pulled back, operating automatic declutch mechanism

Shut-off fin

Discharge valve linkage rod is pushed forward – moving lever to open valve

Shut-off bar is pushed forward by fin

Hose pulled in towards machine

4.16   A stopping system which diverts the flow and declutches the main drive on a hose reel

taken up by each turn of the reel increases as the overall diameter of the pipe coil increases. If it winds at a constant speed then the raingun will gradually move faster as it approaches the reel.

The maximum length of hose is limited more by the pulling force that the tube can withstand than by its capacity to carry the flow of water. The pulling force varies depending on the amount of hose laid out and the crop and soil conditions, and may be as high as 4 tonnes on a 400 m long hose. Hose length may also be limited by the tube diameter. Larger diameter tubes are more rigid than smaller ones and are more difficult to coil. They need larger diameter reels and the amount of hose that can be stored is limited.

When winding in the hose it is important that it is coiled neatly on the reel. On some machines the hose is fed through a fixed hose guide and the hose reel slides back and forth for winding. (figure 4.21). This system ensures that the hose and raingun sledge follow a straight track. On other machines the hose

guide slides back and forth and follows the winds of the coil. With this system there is a tendency for the hose to be pulled back and forth across the travel lane within 7-8 m from the machine.

Care and maintenance
A flexible hose is expensive. Treat it with great care at all times to avoid expensive repair and replacement cost.
1 Although the hose is designed to be pulled across fields, do not pull across roads and similar hard surfaces as this will cause damage.
2 When storing a lay-flat hose for long periods ensure that it is wound on to the reel without twisting and kinking and without undue tension. This can be done by laying out the hose on open ground and slowly winding in the reel using the tractor power take-off (pto) but without applying the tractor brakes. If tension develops in the hose then the tractor will be pulled backwards. This will avoid tight reeling.

4.17 Hose reels can be mounted vertically or horizontally

4.18 A loop is formed behind and pulled along by the machine

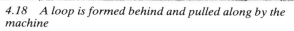

4.19 A wheeled frame lifts and supports the loop clear of the crop

4.20 Coupler for connecting flexible hose to pump and raingun (a) Exploded view (b) Connection to raingun

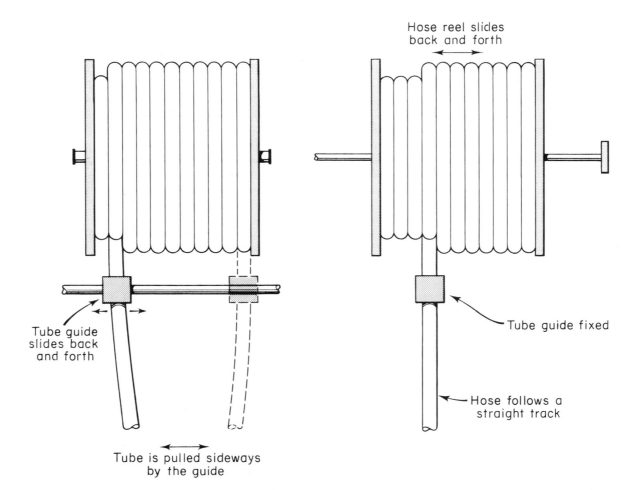

Hose reel slides
back and forth

Tube guide
slides back
and forth

Tube is pulled sideways
by the guide

Tube guide fixed

Hose follows a
straight track

4.21 Two ways of winding in a hose on a hose-reel machine (a) Sliding tube guide (b) Sliding hose-reel

3 Where soils are abrasive (ie containing silica or quartz) hose life can be extended by providing a vegetative cover on the travel lane. Lay-flat hose life can also be increased by turning the hose coupler a quarter-turn clockwise four times during the season. This will expose each section of the hose to the ground for a short period, producing more even wear.

4 When storing hose-reels at the end of the irrigation season, drain out any water left in the hose. This is particularly important where there is the likelihood of heavy frosts. Pull out the hose so that it can be inspected for possible damage. Wind slowly back onto the reel using a tractor power take-off (pto) attachment.

Hose repair

Small cuts or holes in lay flat hose can be repaired easily provided the correct adhesives and parts are available. These normally come in kit form from the manufacturer. Holes are first filled with a mushroom-shaped patch filled with adhesive (figure 4.22). A protective sleeve is then slid along the pipe and glued over the repair.

To repair the more rigid hoses, the damaged part is first removed. The two sections of pipe are then welded by heating the pipe ends and forcing them together. A special welding machine kit is normally supplied by manufacturers.

Small holes in rigid hoses can sometimes be plugged by screwing a self-tapping screw of the right size into the hole.

4.22 Lay flat hose repair. (a) Hole is cleaned out (b) Mushroom shaped repair patch coated with adhesive is folded, pushed into a hole and opened out with 'stem' projecting outwards (c) Repair is weighted during setting using a spacer with hole in centre positioned over roughly trimmed 'stem' (d) Stem is finally trimmed and protective sleeve glued over repair

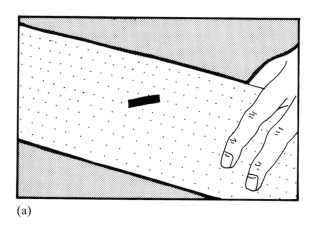

(a)

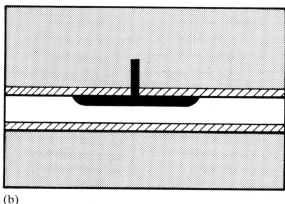

(b)

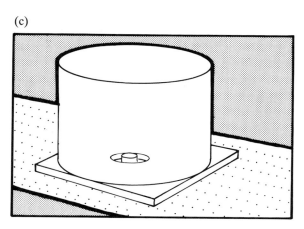

(c)

(d)

Safety

Occasionally the coils of pipe on a hose-reel may need adjusting. Do not attempt to do this while the hose is being pulled out or winding in. The hose is heavy and rigid and fingers or loose clothing could easily be trapped.

4.7 Maintenance

As with all machines regular maintenance is essential if they are to give reliable and efficient service. In irrigation this is especially important as a breakdown at some critical period could have disastrous results for the crop.

Regular weekly inspections should be made to:

1 Ensure that gears are well oiled and chain or pulley belts have the correct tension.
2 Check that all bearings are running smoothly. Of particular importance for hose-reel machines is the main reel bearings.
3 Follow the manufacturers detailed maintenance instructions ensuring that the correct oils, greases and replacement parts are used.

At the end of an irrigation season all water should be drained from the machines and hoses including the water motors. This will avoid corrosion and possible frost damage. It is also a convenient time to inspect the machines thoroughly and replace broken or worn components. Machines should preferably be stored under cover and be raised on stout blocks to lift the wheels off the ground. All bare metal surfaces should be protected against corrosion by a coating of grease and all rubber seals should be removed and stored in a cool dark place (see page 38).

4.8 Field management

Operating conditions
Mobile raingun systems have recently been criticised. The uniformity of water distribution is often said to be poor. The large drops of water produced and applied at high rates can cause considerable damage to crops and soil structure, and cause surface water runoff (figure 4.23). Experience has shown,

4.23 Run off down potato furrows under raingun irrigation

however, that such damage can be reduced and a high degree of uniformity obtained provided rainguns are operated at the recommended pressures, the correct sizes of nozzles are used and the travel lane spacings and alignments are chosen carefully. The continuous slow movement of these machines also helps to provide a more uniform irrigation than might be obtained using conventional static systems.

Raingun systems are suitable for a wide range of crops but care is needed to avoid damage on delicate crops particularly during critical germination and blossom stages. On soils where structure is easily damaged under irrigation (problem known as *capping*) rainguns should only be used after the crop has established a full protective cover. Heavy textured soils can also be a problem as the application rate may exceed the soil infiltration rate and surface water run-off may occur. Where such problems are likely to occur particular attention should be made to the selection of nozzle size to ensure good break-up of the water jet.

Many of the disadvantages associated with rainguns are often out-weighed by the advantages of low cost and low labour requirements.

Travel lanes
Travel lanes have a minimum width of 3 m to provide sufficient space for the raingun carriage and the hose loop in the case of hose-pull machine (figure 4.24).

4.24　Raingun operating along a travel lane

The spacing and direction of travel lanes needs careful selection for good uniformity of irrigation. Lane spacing depends on the throw of the raingun and the direction and speed of the wind. Ideally travel lanes should be at right angles to the prevailing wind. If this is not possible, the effects of wind distortion can be minimised by reducing the spacing between travel lanes. Table 4.1 shows typical spacings in relation to wind speed. Travel direction is also affected by ground slope. If the slopes are too steep then travel direction should be with the main slope. If the travel lane was across the slope, the machine would tend to run off the lane and down the slope (figure 4.25).

If the only way to irrigate a field is by running across the steep slope then it is possible to hold the steel cable or hose in place with steel pins (figure 4.26). These would be lifted out of the ground by a special lifting device as the raingun approached. Properly developed travel lanes, particularly for hose-pull machines can increase both the

efficiency and life of the equipment. When growing tall crops, such as maize or sorghum, the lanes can be prepared and seeded with a low-growing grass. This reduces the drag or force required to pull the traveller along and the wear on the hose. When growing a low crop such as potatoes or lucerne the field can be planted solidly, omitting prepared lanes. Once the crop is established the traveller can forge through the crop creating its own lane during the first irrigation.

Wetted diameter	Wind conditions (m/s)			
(m)	No wind	0-2.5	2.5-5.0	above 5.0
	Travel lane spacing (m)			
60	48	42	36	30
80	64	56	48	40
100	80	70	60	50
120	96	84	72	60

Table 4.1　Travel lane spacings in relation to wind speed.

73

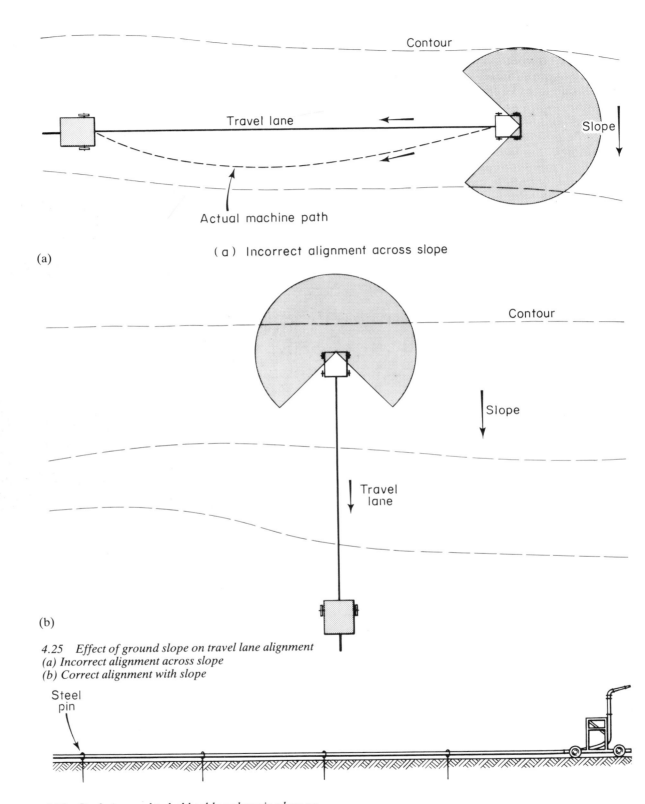

Contour

Travel lane

Slope

Actual machine path

(a)

(a) Incorrect alignment across slope

Contour

Slope

Travel lane

(b)

4.25   Effect of ground slope on travel lane alignment
(a) Incorrect alignment across slope
(b) Correct alignment with slope

Steel pin

4.26   Steel pins used to hold cable or hose in place on
steep cross slopes

Always inspect travel lanes before the season starts for objects which may damage the hose, eg barbed wire, scrap iron, timber, nails, etc. A small amount of time spent doing this will save a lot of trouble later.

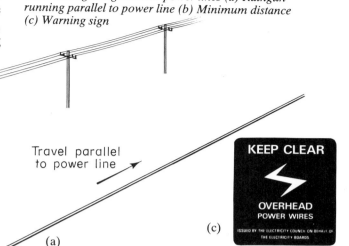

4.27 *Using rainguns near power lines (a) Raingun running parallel to power line (b) Minimum distance (c) Warning sign*

4.9 Safety

1 All moving parts of the machines described such as chains, sprockets, ratchets, etc, are a safety hazard during operation. Take care that limbs and clothes are kept well clear of these moving parts.

2 **Do not** operate a machine until well versed in all its operational procedures.

3 **Do not** operate a machine with any of the guards removed.

4 **Do not** climb on to the hose reel or raingun carriage and make adjustments while the machine is operating.

5 To shut down a machine in an emergency:
   (a) Shut off the water supply at pump or valve.
   (b) Operate the mechanism which normally stops the machine at the end of its run.
   (c) Individual manufacturers build in their own safety devices – read their instructions before using the machine.

4.10 Safety and powerlines

Accidents and even deaths can occur when using rainguns and other irrigation equipment near overhead powerlines. It is not even necessary to actually touch the cables, as arcing can occur to an earthed metal object when close to the lines.

Water jets hitting the lines can also be extremely dangerous. They may cause electrical leakage along the jet to the raingun equipment or cause cables to swing and clash together, resulting in a power failure.

Any raingun operating near a power line must have a minimum nozzle-to-line clearance of 30 m (figure 4.26). This ensures that the solid jet of water is broken up into droplets before reaching the line and avoids electrical leakage and cable clashing. As jet break-up is so important the raingun operating pressure and the spoon drive must be set correctly. The use of ring nozzles instead of taper nozzles will also help considerably.

The raingun travel lane should always be aligned parallel to the powerline and not under it. If this is not possible then the power lines should be moved to a safer location.

5

SPRAYLINES

Spraylines are an alternative to the more common rotary sprinkler. They consist of pipes with small holes or nozzles along their length through which water is sprayed under pressure. There are three types of sprayline:

    Stationary
    Oscillating
    Rotating

5.1 Stationary spraylines

These are sometimes referred to as *perforated pipes*. They consist of light weight portable laterals 50-100 mm diameter connected together using quick couplers similar to those used in conventional systems (figure 5.1). Small holes 1-2 mm diameter are drilled into the top side of the pipe so that water sprays out in all directions wetting a rectangular area. Sometimes small jets or nozzles are screwed into the holes to provide better control over the application of water.

Stationary spraylines operate at pressures between 1.5-5 bar (25-75 lbf/in.^{2}) and can irrigate an area of land 5-15 m wide and up to 200 m long at one setting. The application rates vary from 10-30 mm/h. This depends on the size and spacing of the nozzles. Pressure has very little effect on application rate but does effect the throw of each nozzle and hence the width of the wetted strip.

5.1   Stationary sprayline

When irrigating, the mainline and laterals are laid out and operated in a manner similar to the portable systems using rotary sprinklers. Laterals are moved around the field in the same way. Laterals can be laid directly on the ground or supported above the crop on metal stands approximately 6 m apart.

Stationary spraylines are not normally used for large scale field cropping. They are often used for horticultural crops and are ideal for nurseries rearing small and often delicate seedlings and plants.

Care and maintenance

Many of the care and maintenance procedures outlined for conventional systems also apply to stationary spraylines. Other points to note are:

1 Do not use this type of system under windy conditions. The small water jets are easily distorted. Slight gusts of wind, however, help to redistribute the water drops. Under very still air conditions drops from each jet tend to fall in the same spot and can damage crops and soil.
2 The small nozzles are easily blocked. Careful filtration of the irrigation water is essential to avoid this problem. Periodic flushing of the pipe also helps to stop blockages by removing debris from the lateral.

5.2 Oscillating spraylines

These are also used mainly for small scale horticulture. They consist of lightweight portable laterals 25-50 mm diameter with nozzles 1-2 mm diameter spaced out along the top of the pipe 0.6-1.5 m apart (figure 5.2). The lateral is supported above the crop on stands which allow it to be rotated from side to side through approximately 100° to irrigate a rectangle on either side of the pipe.

Oscillating spraylines operate at pressures similar to the stationary type and irrigate a similar area at one setting.

Pipes are joined together with special rigid couplers so that the whole lateral can be rotated from one place (figure 5.2). Rotation can be done by hand or by an automatic oscillator.

There are two main types:
Water piston oscillator
Water tank oscillator

Water piston oscillator

The water piston oscillator consists of a double-acting piston similar to the type used to drive mobile raingun systems (figure 5.3). Water is supplied from the main pipe line and diverted alternately to each side of the piston by a spring-loaded valve. The piston is connected to a beam clamped to the pipe. As the piston rises and falls the pipe rotates. The speed of oscillation depends on the flow through the water piston.

Water tank oscillator

The water tank oscillator consists of a steel or glass fibre tank of 10-15 litre capacity clamped firmly to the pipe (figure 5.3(b)). It is held in place by a metal frame and a large coil spring. Water from the sprayline is supplied to the tank so that it fills up slowly. As it fills its weight increases and causes the pipe to rotate. Flow to the tank is controlled by a weight-loaded two-way valve. When the tank reaches its lowest point, the weight swings across, cutting off the supply to the tank and allowing it contents to trickle away slowly. As the weight of the tank gradually decreases, the spring brings it back to its original position. The weight on the valve then swings back, opens the flow again and the cycle is repeated.

This is a simple and very reliable oscillator but one complete cycle may take up to 30 minutes. The piston oscillator can work much faster, completing one cycle in less than one minute.

When irrigating, the lateral is connected to the mainline by a length of flexible hose. The pipes can be dismantled for moving, or lifted complete by a group of men and carried over the crop to the next sprinkler position.

Care and maintenance

Many of the care and maintenance procedures outlined for conventional systems apply also to oscillating spraylines. Other points to note are:

1 Windy conditions should be avoided if possible. Effects of wind distortion can be reduced by aligning the line of nozzles more into the wind (figure 5.4).
2 The small nozzles are easily blocked. Filtration of the water supply and periodic lateral flushing are essential to eliminate blockages.

5.2   Oscillating spraylines. (a) General view (b)   Layout   (c) Pipe stand (d) Rigid pipe coupler

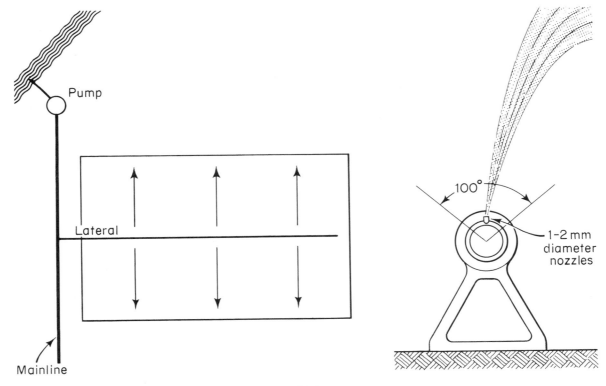

(b)

(c)

(d)

(b)

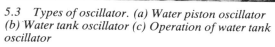

*5.3   Types of oscillator. (a) Water piston oscillator
(b) Water tank oscillator (c) Operation of water tank
oscillator*

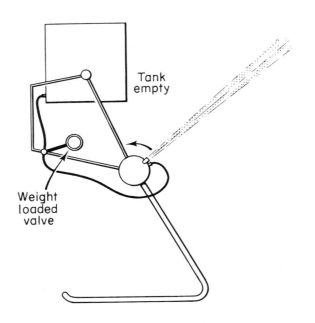

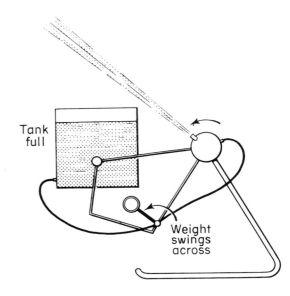

(c) Operation of water tank oscillator

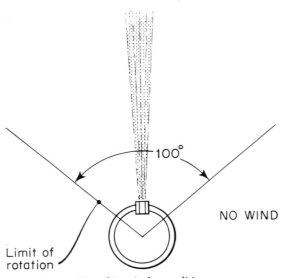

NO WIND

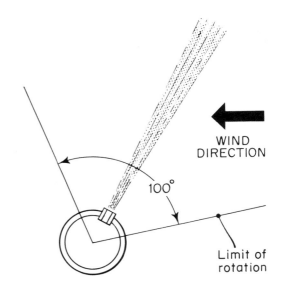

WIND DIRECTION

5.4 *Aligning lateral in windy conditions*

3 Use only filtered water for the oscillators. Grit or sediment in the water may block pipes and valves and cause undue wear on the piston and cylinder. This will result in a loss of power from the oscillator.

4 Grease all moving parts on the oscillators every 50 working hours.

5.3 Rotating spraylines

These are sometimes called *rotary irrigators* or *boom sprinklers* (figure 5.5). The main feature is a pipe boom with an overall length up to 80 m pivoted at the centre. It has nozzles varying from 4-8 mm diameter spaced out along its length to provide an even distribution of water as it rotates. A range nozzle fitted into each end of the boom increases the area that can be irrigated at one setting. The boom is mounted on a wheeled carriage powered by an internal combustion engine or on a tractor. It is supported from the centre by steel cables. Water is supplied from the mainline through a swivel joint to the boom and nozzles.

5.5   Rotating sprayline

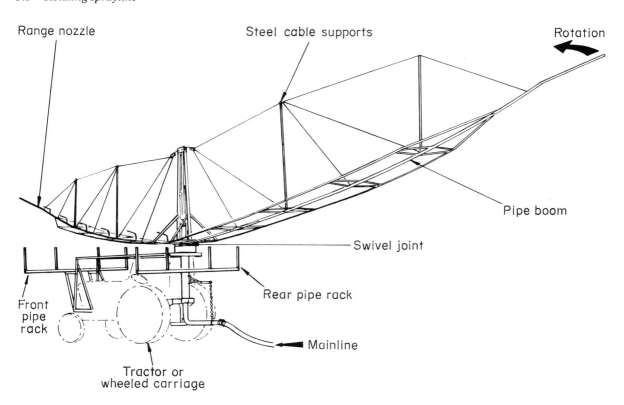

Range nozzle

Steel cable supports

Rotation

Pipe boom

Swivel joint

Front pipe rack

Rear pipe rack

Mainline

Tractor or wheeled carriage

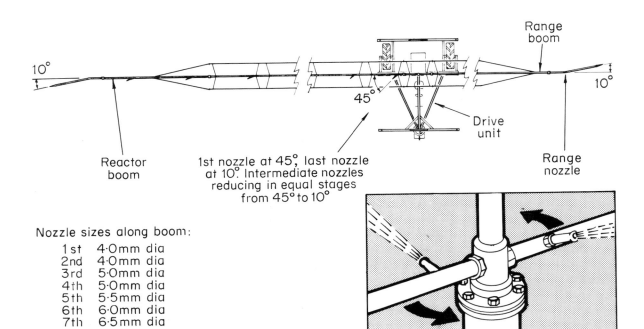

10°   45°   Range boom   10°

Reactor boom

1st nozzle at 45°, last nozzle at 10°. Intermediate nozzles reducing in equal stages from 45° to 10°

Drive unit

Range nozzle

Nozzle sizes along boom:
  1st   4·0mm dia
  2nd   4·0mm dia
  3rd   5·0mm dia
  4th   5·0mm dia
  5th   5·5mm dia
  6th   6·0mm dia
  7th   6·5mm dia
  8th   6·5mm dia
  9th   ——
  Last 8·0mm dia

(a) Typical layout of nozzles

(b) Rotation caused by back pressure from nozzles

5.6 *Arrangement of nozzles on a typical rotating lateral (a) Typical layout of nozzles (b) Rotation caused by back pressure from nozzles*

5.7 *Typical layout for rotating sprayline system*

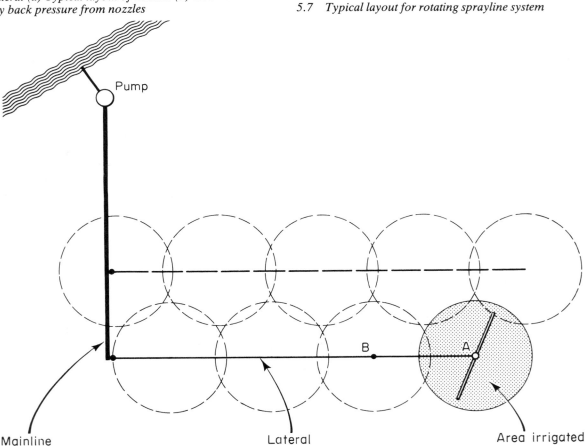

Pump

Mainline

Lateral

B   A

Area irrigated

Rotating spraylines operate at pressures from 5-6 bar (75-90 lbf/in.^{2}) with discharges from 40-75 m/h. An area up to 1.2 ha (a circle 125 m diameter) can be irrigated at one setting of the machine. The application rate is usually very low, around 7-10 mm/h.

Rotation of the boom occurs as water flows from the nozzles under pressure. This creates a back-pressure or pushing action in the opposite direction to the flow and causes the boom to rotate. No other external drive is necessary. The nozzles are aligned specially to take full advantage of this effect (figure 5.6). One revolution takes 1-2 minutes.

A typical layout for a rotating sprayline is shown in figure 5.7. A permanent or portable mainline is laid along the edge of the field from the pumping station. The portable lateral pipes which supply water to the machine are stored on pipe racks under the boom (figure 5.8). As the machine is towed across the field to its first position (A) the pipes are gradually off-loaded and coupled together. This is a two-man job, one man driving the tractor and one man off-loading the pipes. On reaching position (A) the machine is coupled to the lateral. A special telescopic coupling is used to make it easier to position the machine (figure 5.8).

Irrigation is started by slowly turning on the water supply. Once the pressure builds up, the boom automatically starts to rotate. A typical application of 50 mm of water takes from 5-7 hours depending on the application rate.

At the end of an irrigation, the water supply is shut down and the machine moved to the next position (B). The lateral pipes no longer required are loaded back on to the machine pipes racks. Such a move can take as little as 10 minutes. This procedure is repeated until the machine arrives back at the mainline. It is then moved across to the next line. This operation takes approximately 30 minutes. It is better to offset the position of the machine along this line and use a triangular spacing. This greatly improves the uniformity of water distribution.

Care and maintenance
Many of the care and maintenance procedures outlined for conventional systems apply also to rotating spraylines.

Other points to note are:
1 When moving the machine from one part of a field to another it is not necessary to dismantle it. Align the boom in the direction of travel with one man holding the lead end in place with a rope (figure 5.9). This man can rotate the boom to move the machine around obstacles such as farm buildings.
2 When moving over sloping ground it is better to drive the machine with the slope rather than across it.

At the end of the irrigation season:
1 The boom structure can usually be dismantled for easy storage.

5.8 Rotating sprayline details. (a) Pipe racks mounted under boom (b) Telescopic coupling between machine and lateral makes it easier to position machine

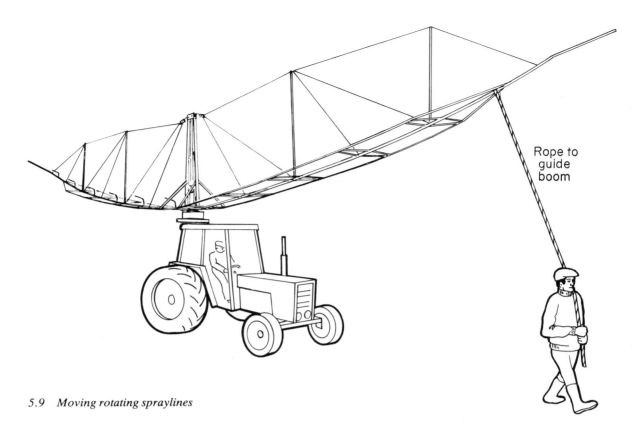

Rope to
guide
boom

5.9   Moving rotating spraylines

2 Take care when dismantling the boom arms that the water seals are removed, cleaned and stored properly.
3 Lubricate all moving parts, particularly the swivel joints.
4 Replace any items showing signs of wear. This is best done during the winter period so that no time is lost through breakdowns early in the next season.

Safety and powerlines
Because this type of machine is rather large and somewhat cumbersom to move, special care is needed when operating close to powerlines, see page 75. A minimum distance of 15 m is essential between the end range nozzle and the cables and the travel path should never pass under the powerline.

6

MOBILE LATERAL SYSTEMS

Many new irrigation systems have been developed in recent years to try and combine the advantages of conventional sprinkler systems with the mobility of rainguns. These are called *mobile lateral systems* because they use laterals which move continuously while applying water. They were developed because:

1 Conventional systems rely very much on labour to move equipment. In many areas labour is difficult to find or is too expensive.
2 Mobile rainguns are sometimes unsuited to the more delicate crops and areas of land where soil structure is easily damaged under irrigation.

There are two main types of mobile-lateral system:

Centre-pivot systems

Side-move systems

6.1 Centre-pivot systems

These machines consist of a single galvanised steel lateral which rotates in a circle about a fixed point in the centre of the field (figure 6.1). The lateral is supported above the crop with as much as 3 m ground clearance on A-shaped steel frames using cables or trusses. The frames are spaced approximately 30 m apart and are mounted on wheels or crawler tracks. Laterals vary in length from 150-600 m.

A typical field layout is shown in figure 6.2. Water is suppled to the centre pivot by a mainline laid across the field or from a well located near the pivot. Water flows through a swivel joint to the lateral and sprinklers. When irrigating, the lateral rotates continuously about the pivot, wetting a circular area up to 100 ha depending on the length of the lateral. One revolution can take from 1-100 hours depending on the amount of water to be applied. The slower the lateral rotates the more water is applied.

Water application
Small rotary sprinklers are normally used on centre-pivots and so they operate at pressures similar to conventional systems. As the lateral moves in a circle, special provisions are made to obtain an even distribution of water. This is because sprinklers near the pivot irrigate a much smaller area than those near the outer end. Uniform watering is achieved by gradually increasing the application rate towards the outer end of the lateral. This is done in two ways:

Varying the size of sprinklers and
Varying the spacing of sprinklers.

The first method uses sprinklers of different sizes equally spaced along the lateral. Small sprinklers are used close to the pivot and larger ones are used towards the outer end (figure 6.3).

The second method uses the same size of sprinkler throughout but varies the spacing along the lateral. Sprinklers are placed closer together towards the outer end. This method can make maintenance simpler as all the sprinklers are the same size and require the same spare parts.

Typical application rates vary from 5 mm/h

6.1 *Centre-pivot machine (a) General view (b) Aerial view (c) Centre-pivot*

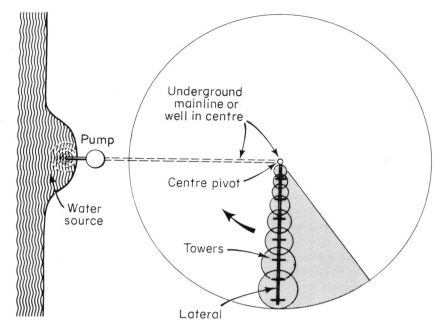

6.2 *Typical field layout for centre-pivot machine*

6.3 *Ways of achieving a uniform water application. (a) Varying size of sprinklers (b) Varying spacing of sprinklers*

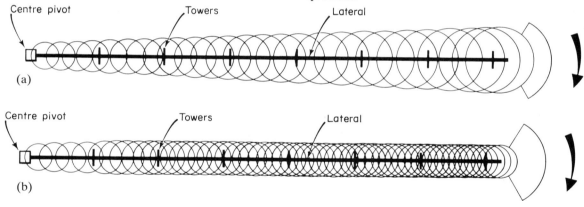

near the pivot to more than 30 mm/h at the outer end.

Drive mechanisms

The centre-pivot moves continuously when irrigating. Each tower has its own independent drive unit. This allows each tower to travel at a different speed. The distance around the circular area increases further away from the centre (figure 6.4) and so each tower must travel faster if the lateral is to rotate in a straight line.

Power to drive the lateral can be hydraulic or electrical. Hydraulic power is supplied by water pressure from the lateral and is used to drive water motors similar to those used on

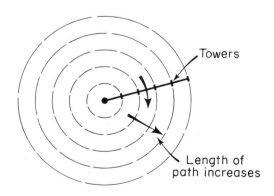

6.4 *Centre pivot tower paths*

mobile rainguns see page 61. Sometimes a separate water supply or oil hydraulic system is used. The speed of each tower and hence the lateral is controlled by the flow to the water motors. The lateral is kept in a straight line by a system of control cables along the full length of the lateral. If, for example, one tower starts moving ahead of the others and pulls the lateral out of line, a cable tightens, closes the supply valve to the water motor and slows it down again. Similarly, if one of the towers lags behind the others, a second cable tightens and opens the supply valve and increases the speed of the tower.

The rotating speed of the lateral is controlled by regulating the speed of the end tower. As all the other towers are linked by the control cables they will each automatically respond as the end tower starts to move.

The main disadvantage of using water motors powered from water pressure in the lateral is that the system will only move when irrigating. It is not possible to move the lateral when dry.

When an electric power supply is available each tower is fitted with its own electrical motor, (usually from 0.5-1.5 HP) (figure 6.5). This drives the wheels or crawler tracks through a gear box or chain drive. The rotating speed of the lateral is controlled by regulating the speed of the end tower from a control panel close to the centre-pivot. The lateral is kept in line in a similar manner to the hydraulic drives. The control cables are used to switch individual electric motors on and off.

Field management

Centre-pivots operate best on sandy soils which absorb water quickly and can support the heavy wheel loads from the towers. They can be used on most crops grown on flat and undulating ground (figure 6.6). Special flexible couplings are used to connect spans together to allow for the changes in alignment.

One main advantage of this system is that it can be fully automated and controlled from a panel near the pivot or remotely from some office nearby (figure 6.7). Time clocks are used to start and stop the machine and many safety devices are used for protection. For example, if the water pressure suddenly drops or one of the tower drives breaks down, the system will automatically stop irrigating and an alarm will alert the operator. Several centre-pivots covering very large areas can easily be controlled thus and maintained by a few men. They would, however, need to be highly skilled and have experience in using these machines.

One main disadvantage is that it irrigates only circular areas and not the corners of square fields (figure 6.2). This can be overcome by using special booms or a large raingun at the outer end of the lateral, automatically set to operate only when travelling past the corners (figure 6.8).

6.6 On undulating ground special flexible couplings are used to allow changes in alignment.
(a) Junction between spans (b) Flexible coupling

6.5   Electric tower drive

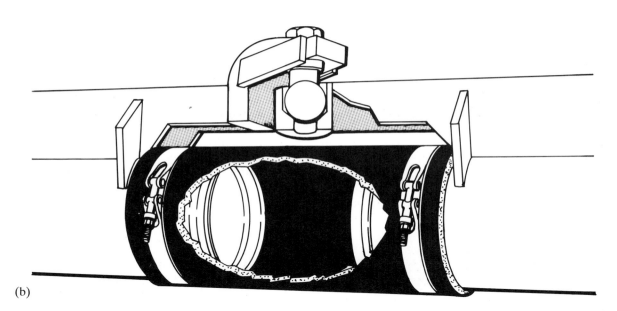

(b)

6.7　*Control panel for a centre pivot*

6.2 Side-move systems

These are systems which are designed to irrigate square or rectangular fields while continuously moving.

One system for irrigating large fields is similar in construction to the centre-pivot, supporting a lateral over the crop on towers, (figure 6.9). Water is supplied by a flexible hose or from a small canal along the side or centre of the field.

The lateral is powered by water motors on each tower and is controlled in the same ways as a centre-pivot. Electric motors are not normally used because of the difficulty of supplying power to the moving lateral. A steel guide cable is laid out along the ground from the machine and anchored at the opposite end of the field. This keeps the lateral moving in the right direction. A winch powered by a water motor gradually winds in the cable.

Field management

This system can be automated in the same way as a centre-pivot but when the lateral reaches the far end of the field it has to be moved manually back to the beginning. This means moving a heavy machine over newly irrigated land. On sandy soils this may not be a problem but on clay soils the towers may sink into the soil even when crawler tracks are used. It may be necessary to wait a few days

6.8　*Corner unit for a centre-pivot*

6.9 *Side move system (a) General view of lateral*

until the soil surface dries out. Another way is to divide the field into two parts for irrigation (figure 6.9). Irrigation starts on line A-B and the lateral moves to the centre of the field C-D. Irrigation then stops and the lateral is moved dry to the other end of the field (E-F). Irrigation starts again moving back towards C-D. On large fields this may take several

(b) Water supplied from a small canal down centre of field

days. When C-D is reached the first half of the field should have dried enough for the machine to move dry to start the next irrigation from A-B.

Another side-move system is based on machines originally built for rainguns (figure 6.10(a)) small rotary sprinklers are positioned along booms fitted to each side of a portable carriage. A typical layout is shown in figure 6.11. When irrigating the carriage is positioned at one end of the field and slowly pulled across using a steel guide cable and winch in the same way as the hose-pull raingun. A strip of land up to 70 m wide and 400 m long can be irrigated at one setting (2.8 ha).

Water is supplied by a flexible hose pulled along behind the carriage or by a more rigid hose coiled on the carriage (figure 6.10(b)). The rigid hose is used only to supply water and not to pull the machine as with the hose-reel raingun, as it is far too heavy.

Some manufacturers now supply mobile rainguns that can be easily converted to booms if required thus adding to the flexibility of the system.

6.3 Care and maintenance

Many of the comments that can be made about care and maintenance of mobile lateral systems have already been made in detail in chapters 3 and 4.

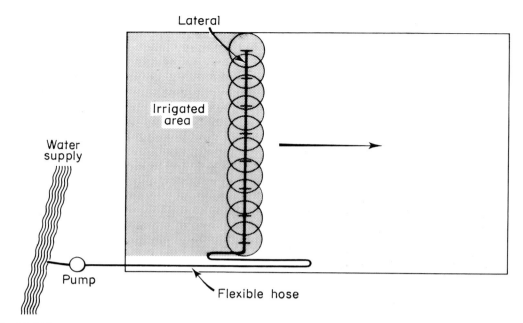

(c) *Typical field layout*

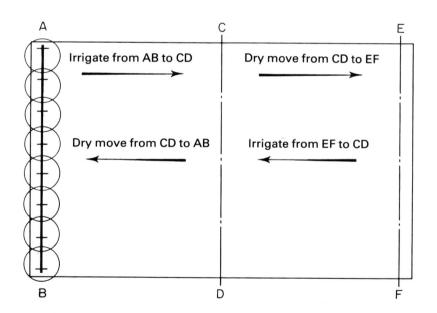

(d) *A method of irrigating clay soils*

6.10   *Small side-move systems*
(a) Flexible hose pulled along behind machine
(b)Flexible hose coiled on machine

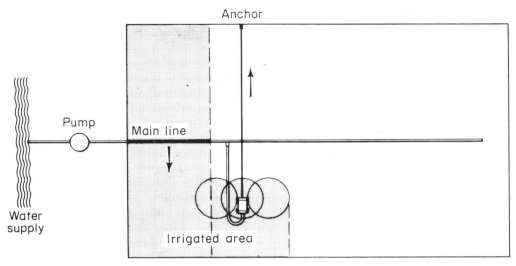

6.11 *Typical field layout for a small side-move system.*

6.4 New developments

An interesting new development on mobile lateral systems is the replacement of small rotary sprinklers with fixed spray nozzles (figure 6.12). There are two main reasons for this:

1 Less energy is required to operate the system. Fixed nozzles operate at much lower pressures than rotary sprinklers; from 0.3-1.0 bar (5-15 lbf/in.^{2}).

2 Water distribution is improved. Fixed nozzles spray down towards the ground rather than up in the air. This means that spray is not easily blown about by strong winds. Some machines use nozzles suspended from long tubes to spray water in the crop below the canopy.

The main objection to fixed nozzles is that the area irrigated at one time is much smaller than with rotary sprinklers (Figure 6.12(c)). If a field is to be irrigated in the same time period then a machine using fixed spray nozzles must move faster and apply water at much higher rates: usually between 100-300 mm/h. This is much higher than most soils can absorb over long periods. To prevent surface water run off, cultivation practices can be used which pond the water on the surface close to the plants. Such practices as tied ridging on row crops, a common technique for soil conservation, are being examined for this purpose.

6.12 Low pressure sprinkler systems (a) Fixed spray nozzles on a side-move system (b) Nozzle detail (c) Reduced are of coverage

(b)

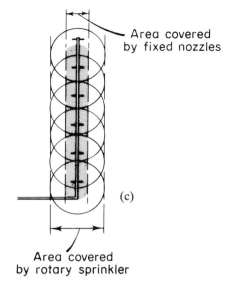

Area covered
by fixed nozzles

(c)

Area covered
by rotary sprinkler

7

CHOOSING A SPRINKLER SYSTEM

Many factors must be considered when deciding which is the best sprinkler system to use. These include:

Land topography

Field shape

Soils

Crops

Labour

7.1 Land topography

Sprinklers can be used on hilly and uneven land which is unsuitable for surface flooding methods. The type of system used depends on the land slope. This is a measure of the difference in ground elevation between two places in a field (figure 7.1). It is referred to as *per cent.* (%). That is the number of metres difference in elevation for 100 m of horizontal distance.

If land slope is less than 5% any type of system can be used. On land steeper than this it becomes more difficult to keep mobile rainguns in line when moving across the slope. Slopes greater than 15% are only suitable for conventional portable, permanent and semi-permanent systems. There could also be soil erosion and soils would need careful protection under sprinkler irrigation.

Land which is undulating and irregular can present problems for most systems. Conventional systems can be designed to fit most conditions. It may be difficult to use mobile rainguns and mobile laterals as they require reasonably smooth paths to move along. Flexible couplings can be used on mobile laterals to allow them to flex as they travel over high and low ground.

7.1 Calculating land slope

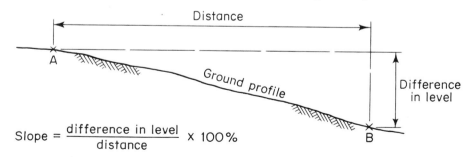

$$\text{Slope} = \frac{\text{difference in level}}{\text{distance}} \times 100\%$$

Example

From a plan of a field the ground level at A is 50.0 m and B is 45.0 m. If the distance from A to B is 150 m, what is the land slope?

$$\text{Slope} = \frac{50 - 45}{150} \times 100\%$$

$$= 3.3\%.$$

7.2 Field shape

All the systems described are easily adapted to regular shaped fields such as squares or rectangles. Some however, are specifically designed only for regular fields such as the larger side-move systems.

Conventional, mobile rainguns and small side-move systems can be adapted to irregularly shaped fields. Centre-pivots only irrigated circular areas and as much as 20-25% of a square field will be lost unless a special corner irrigating device is available.

7.3 Soils

A sprinkler system must always be adapted to the soil conditions so that the application rate is always less than the rate at which the soil can absorb it. This avoids surface water run off and soil erosion. For example if the soil is a clay loam and only absorbs water slowly then a sprinkler system which only applies water slowly must be chosen.

If the soil surface structure is easily damaged under irrigation (capping) then mobile rainguns should not be used.

7.4 Crops

Most crops can be irrigated using conventional systems including field, pasture and row crops, vineyards and orchards.

Mobile rainguns are also suitable for most crops. They may, however, cause damage to young plants and delicate crops such as tomatoes.

Some side-move systems can only be used on crops less than 1.5 m high. Others including centre-pivots and rotating spraylines can operate on crops over 3 m high. These machines are not usually used on orchards or vineyards unless trees are small enough and special paths are provided for the wheels.

7.5 Labour

Labour is required to operate the irrigation systems described. The cost of hiring labour and its availability may affect the choice of system. Conventional portable systems require large labour groups. Other systems are designed to reduce labour needs to a minimum. These include conventional permanent and semi-permanent systems, mobile rainguns and mobile lateral systems. Many of these rely on machinery and so the few men that are required must be highly skilled in operation and maintenance.

8

PUMPING PLANT

A pump is a machine which changes mechanical energy produced by an internal combustion engine or electric motor into useful water energy. In sprinkler irrigation this energy provides the pressures and discharges needed to distribute water in the mainline and laterals to the sprinklers.

8.1 Centrifugal pumps

There are many different types of pump available to suit a wide range of tasks. The most common type used in sprinkler irrigation

is the centrifugal pump. It is best suited to the pressure and discharge requirements of sprinklers, basically simple in design, easy to use and relatively inexpensive to buy and maintain.

How it works
To understand how a centrifugal pump works, consider first how centrifugal forces occur. Most students will at some time have spun a bucket of water around at arm's length and observed that no water falls from the bucket even when it is upside down (figure 8.1).

8.1   Water is held in bucket by centrifugal force

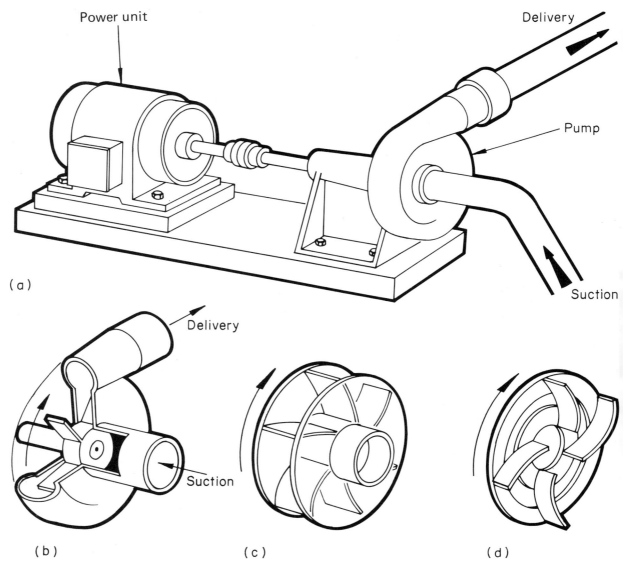

(a)

(b)   (c)   (d)

8.2 Centrifugal pump details. (a) General arrangement (b) Pump scroll (c) Closed impeller (d) Open impeller (e) Section through portable pump showing impeller gear box and seals

Water is held in the bucket by centrifugal forces created by spinning the bucket. If a small pipe is now fixed to the bottom of the bucket, water will be discharged through it as the bucket spins. The faster it spins, the more water will be discharged.

A centrifugal pump makes use of this idea and can be thought of as many 'buckets' all spinning around together. The 'buckets' are replaced by an impeller, with many blades or vanes, which spins at high speed inside a scroll-shaped casing (figure 8.2).

Water is drawn into the pump from the source of supply through a short length of pipe. This is called the *suction*. As the impeller spins, water is thrown outwards by the impeller vanes, collected by the pump casing and guided towards the outlet. This is called the *delivery*.

Impellers
Some simple pumps have straight impeller vanes. These tend to be inefficient but are cheap to make and are often used for small tasks where efficiency is not important. Most irrigation pumps have vanes which curve backwards. Water enters and leaves the impeller more smoothly. Less energy is lost

from turbulence and the pump operates more efficiently.

Some impellers have side plates or shrouds and are called *closed impellers*. When the water contains debris or solids, *open impellers* are used to reduce the risk of blockage. Impellers are made by casting from iron or bronze.

Casings
Pump casings are made by casting from iron. They are normally made in two parts which are bolted together. This enables the casing to be dismantled easily when the impeller requires maintenance or replacing.

Seals
Centrifugal pumps use two types of seal:
  Water seals
  Mechanical seals
Figure 8.2 shows how these seals are used on a typical closed impeller pump.

Water seals prevent water leaking from one part of a pump to another. Two seals B-B are fitted to either side of the impeller to keep water on the delivery side separate from the suction side. During use these seals tend to wear. Sometimes brass wearing rings are fitted to the impeller so that only the rings need to be replaced and not the whole impeller. The water seals are water lubricated. No oil is required. The pump should not be run when it is empty as this will cause very rapid wear on the seals.

Mechanical seals A separate oil lubrication from water in the pump casing. The pump drive shaft rotates in a bearing which requires oil lubrication. To stop water leaking into the bearing a rubber or neoprene ring is fitted between the casing and the drive shaft. It can be held in place by a spring.

Size
Pumps are often described by the diameter of the delivery pipe connection, for example, a 100 mm pump or a 150 mm pump. Table 8.1 is a guide to selecting suitable pump sizes for different flows. This is only a guide and should not be used for design.

Pump size (mm)	Discharge m³/h
50	30-60
75	60-100
100	100-140
125	140-180
150	180-220

Table 8.1 A guide to selecting pumps

8.2 Types of installation
Pumps may be permanently installed or mobile.

Permanent
Permanent pumps are normally used with solid-set and permanent sprinkler systems or where a permanent mainline has been

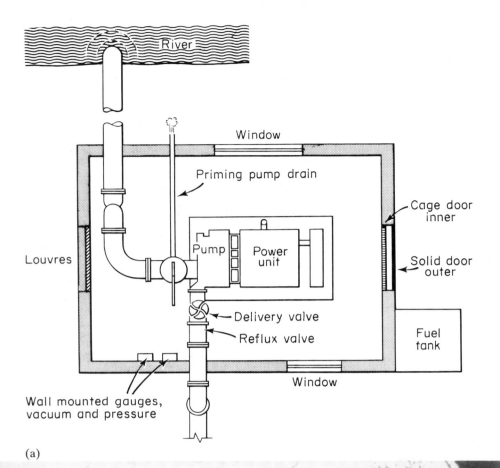

River

Window

Priming pump drain

Cage door
inner

Pump Power
unit

Solid door
outer

Louvres

Delivery valve

Reflux valve

Fuel
tank

Wall mounted gauges,
vacuum and pressure

Window

(a)

(b)

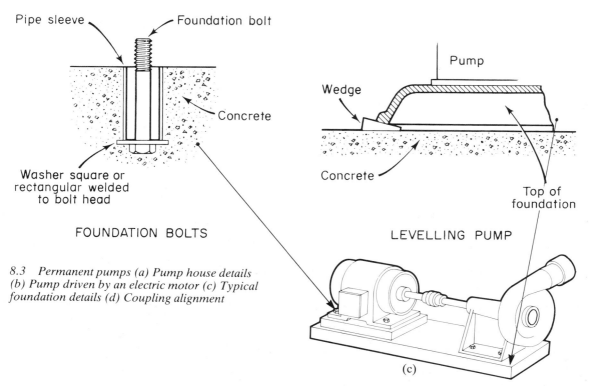

Pipe sleeve — Foundation bolt

Concrete

Washer square or
rectangular welded
to bolt head

FOUNDATION BOLTS

Wedge

Pump

Concrete

Top of
foundation

LEVELLING PUMP

*8.3 Permanent pumps (a) Pump house details
(b) Pump driven by an electric motor (c) Typical
foundation details (d) Coupling alignment*

(c)

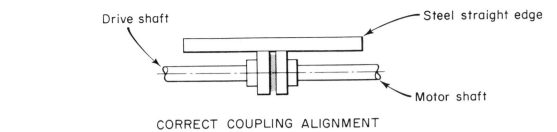

CORRECT COUPLING ALIGNMENT

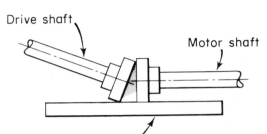

INCORRECT ANGULAR ALIGNMENT

(d)

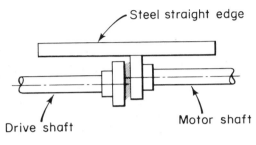

INCORRECT PARALLEL ALIGNMENT

installed. The pump is coupled to an electric motor or internal combustion engine. The whole unit is usually bolted to a strong concrete floor inside a building to protect it from dust and the weather (figure 8.3). Such a building also provides an ideal cover for repair and maintenance work even in the poorest weather conditions.

Some pumps are supplied complete with power unit on a single steel base plate. These can be secured using bolts which are grouted into the concrete slab. Care is needed to make sure the base plate is level. Steel wedges placed between the base and the concrete can be used for fine level adjustments.

When pumps and their power units are supplied separately, each unit must not only be secured to the concrete base, but aligned correctly so that the two can be coupled together properly. In such cases flexible couplings are used. Correct coupling alignment can be checked using a steel straight edge.

Mobile
Mobile pumps are normally used with portable irrigation systems. They are often used where supplementary irrigation is practised when equipment may be needed at several locations around the farm.

Mobile pumps are either mounted on a wheeled trolley or supported on a 3-point linkage on a tractor (figure 8.4). They can be driven by an internal combustion engine coupled directly to the pump or by a tractor pto drive. Electric motors are not normally used on mobile pumps because of the difficulties in providing a supply of electricity. Pumps should always be located on firm level ground and held securely in position during the period of operation.

Trolley mounted pumps which are driven by a tractor power take-off (pto) should be placed straight behind the tractor so that the shaft is as straight as possible. Most pumps are designed to work only when rotating in a clockwise direction. It is useful to check that the tractor pto also rotates clockwise as some do drive anticlockwise.

Pumps mounted on a tractor 3-point linkage are usually driven by the pto drive. It should be positioned so that the shaft is as straight as possible with no lateral or vertical movement when running. Adjustments should also be made to the 3-point linkage so that the pump

8.4 Mobile pumps. (a) Pump mounted on a wheeled trolley (b) Pump supported on tractor 3 point linkage (c) Coupling detail for pumps mounted on 3-point linkage (d) Coupling detail for trolley-mounted pumps (e) Diesel engine coupled directly to pump

unit does not drop when operating. If it drops, the drive shaft alignment will change.

When installing permanent and mobile pumps:
1 Align the pump naturally with the surrounding pipework.
2 Do not force pipes into place as this may uspset the alignment of the pump and the power unit.
3 Support pipes independently of the pump to avoid undue stress on the pump casing.

8.3 Suction

Although some centrifugal pumps are located below the surface of the water supply, in many instances they are installed above (figure 8.5). In such cases, water has to be sucked or lifted up a short length of pipe into the pump. The difference in height between the water surface and the pump is called the *suction lift*.

When a pump is operating it draws water in much the same way as a person sucks water through a drinking straw. There is a limit to how high water can be lifted in this way and it depends on atmospheric pressure. This is the pressure that the atmosphere exerts on the earth's surface and on the surface of the water supply, and is approximately 1.0 bar or 10 m head of water. By sucking, low pressure

(b)

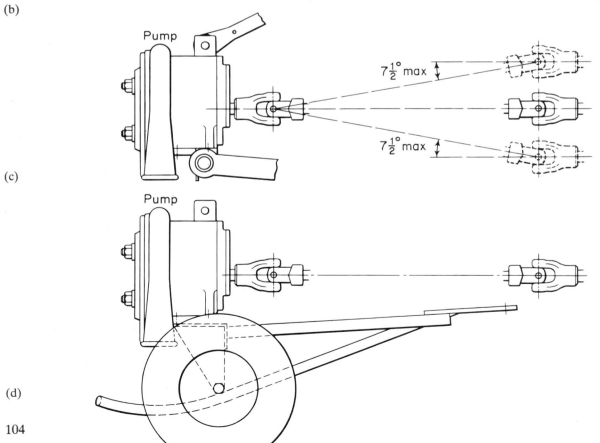

(c)

Pump

$7\frac{1}{2}°$ max

$7\frac{1}{2}°$ max

(d)

Pump

(e)

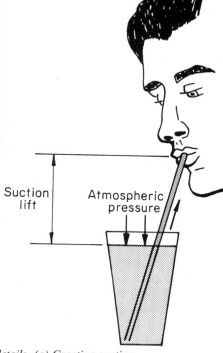

Suction lift   Atmospheric pressure

occurs in the drinking straw. Atmospheric pressure pushes down on the water surface and forces water up the tube. As atmospheric pressure is the main driving force, this puts a practical limit on the height to which water can be lifted in this way. Ideally this would be 10 m but because of friction in the pipe and pump the limit is normally about 7 m. If the reader were to try and drink using a straw more than 7 m long he would find great difficulty in sucking sufficient water to satisfy his thirst. If the straw was 10 m or more he would find his task impossible! Similarly if the suction lift on a pump exceeds 7 m there will be a large fall in output and in some cases the flow may stop completely.

8.5   Pump suction details. (a) Creating suction (b) Pump located below supply water level (c) Pump located above supply water level (d) Suction components

Components

The suction side of a pump is made up of several components (figure 8.5(d)).

The suction pipe connects the water supply to the pump. During pumping the pressure inside this pipe is lower than the surrounding atmosphere and so it must be made from rigid materials that will not collapse inwards.

A *reflux valve* and *strainer* are fastened to the inlet of the suction pipe. The reflux valve is sometimes called a *check* valve or *non-return* valve. It only allows water to flow one way into the pipe. When the pump is operating the flap opens drawing in water. Whe the pump

stops the flap closes keeping the pump and suction pipe full of water. This makes it easier to restart the pump. The reflux valve also contains a strainer grill to stop debris from being sucked into the system. Where there is a lot of debris of floating weeds a finer strainer may be needed.

As there is no outward pressure in the suction pipe, special pipe couplings are needed. In mainlines and laterals some couplings use water pressure to seal joints and stop water leaking out. In suction pipes the problem is quite different and the couplings are sealed to stop air leaking into the pipe and depriming the pump (see Priming below). One method of joining the pipes is by sliding them into a coupling sleeve and pushing tapered rubber seals into the gap between the pipe and sleeve. The seals are expanded in the gap using clamping rings which are drawn together by bolts. To ensure an even pressure

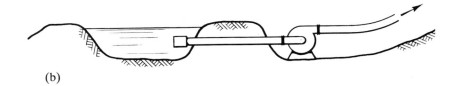

(b)

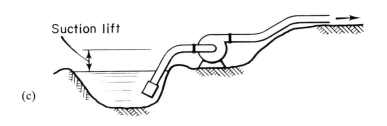

Suction lift

(c)

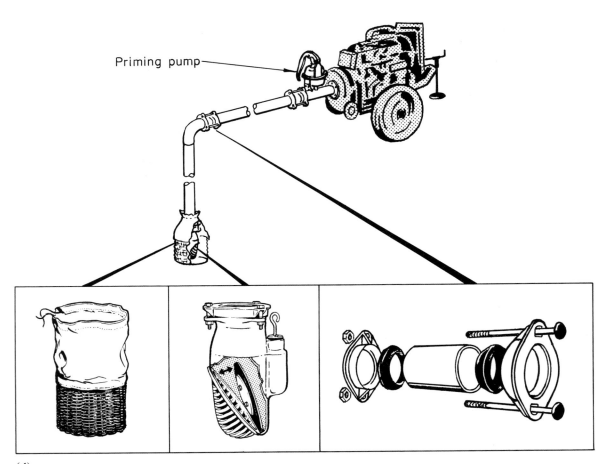

Priming pump

(d)

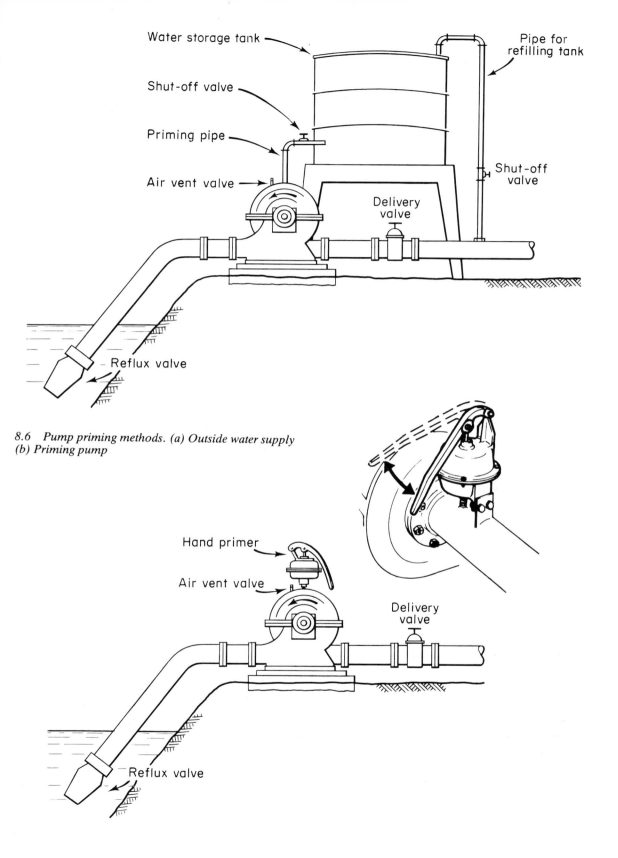

Water storage tank

Pipe for refilling tank

Shut-off valve

Priming pipe

Shut-off valve

Air vent valve

Delivery valve

Reflux valve

8.6   *Pump priming methods. (a) Outside water supply*
(b) Priming pump

Hand primer

Air vent valve

Delivery valve

Reflux valve

107

all around the seals, the bolt are tightened evenly in turn, a little at a time. Another method is to use mechanical seal couplers as shown in figure 3.20.

Priming

Centrifugal pumps will only draw water when the pump casing and suction pipe are full of water and all the air has been removed. Removing air and filling the pump and suction with water is called *priming*. There are several ways of doing this, the most common methods use:

An outside water supply

A priming pump

Outside water supply

An outside water supply may be a tank of water placed close to the pump which has sufficient capacity to fill both the suction pipe and the pump with water (figure 8.6(a)). To prime the pump, the delivery valve is first closed. Water is then allowed to flow from the tank into the pump. Outflow is prevented by the reflux valve. Air is forced out through a small vent on top of the pump. When the suction pipe and pump are full of water, the pump is started and the delivery valve slowly opened to discharge water into the mainline. The tank is refilled from a small pipe on the delivery side of the pump during operation.

Priming pump

A priming pump is normally a simple hand-operated air pump connected to the suction pipe close to the centrifugal pump (figure 8.6(b)). A handle is used to move a diaphragm in the priming pump chamber. Air is drawn into the chamber from the centrifugal pump and suction pipe on the 'upstroke' and discharged through an outlet valve on the 'downstroke'. To prime the pump, the delivery valve is first closed. All the air is then sucked out of the pump and suction pipe until water flows from the priming pump. A reflux valve on the end of the suction pipe prevents water from flowing back out of the pump. The pump can then be started and the delivery valve slowly opened to discharge flow into the mainline.

If the pump is located below the supply water level there is no need for such priming devices. Water will flow under gravity to fill the pump and suction pipe.

Installation

When installing a pump suction pipe and fittings:

1 Avoid high suction lifts by locating the pump close to the supply water level. But, make sure the pump is on firm level ground and not at risk from flooding (figure 8.7).

2 Lay pipe-work on a uniform upward slope. Avoid high spots where air can collect. This will reduce the pump efficiency and may cause depriming.

3 Place the inlet of the suction pipe at least 0.6 m below the lowest water level. This stops air being drawn into the pipe through a whirlpool or vortex.

4 Place the inlet of the suction pipe well above the bed of a stream to avoid sucking in stones or floating weed.

5 If insufficient water depth is available use an empty oil drum or barrel sunk into the bed of the stream to provide a sump.

6 Avoid short pipe bends close to the pump. They will disturb the flow of water causing noisy operation and loss of efficiency. If a bend is necessary keep it well away from the pump (figure 8.5(d)).

8.4 Delivery

The delivery side of a pump comprises pipes and fittings to connect the pump with the mainline (figure 8.8).

A *delivery* valve is connected to the pump outlet. This controls the discharge and pressure in the mainline. It is closed before starting so that the pump can be primed. Once the pump is running it is slowly opened to deliver the flow.

A reflux valve is connected downstream of the delivery valve. This allows water to flow one way only out of the pump into the mainline. When a pump stops, water in the mainline can flow back towards the pump causing a build up in pressure. (See 2.16 Waterhammer.) This can damage the pump casing and possibly the power unit. The reflux valve prevents this by closing and stopping this flow from reaching the pump.

A pressure gauge is normally fitted to the pipe between the pump and the delivery valve. When starting up the operator can use this to check that the pump is primed and working properly. It can then be used to

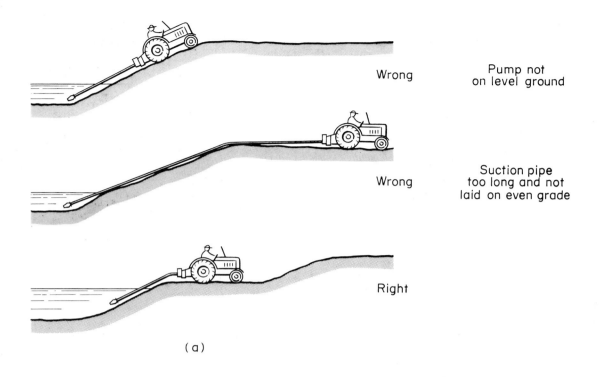

Wrong — Pump not on level ground

Wrong — Suction pipe too long and not laid on even grade

Right

(a)

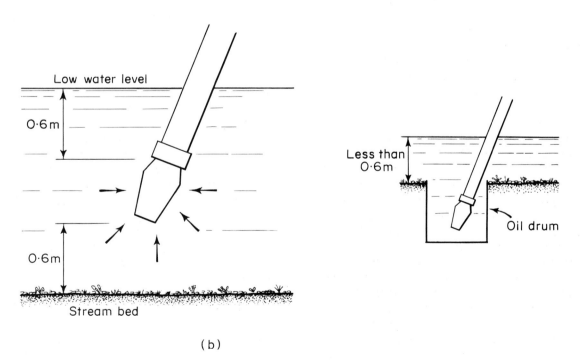

Low water level

0·6 m

0·6 m

Stream bed

Less than 0·6 m

Oil drum

(b)

8.7 *Installation of pump suction*
(a) Locating pump and suction pipe
(b) Siting the suction inlet

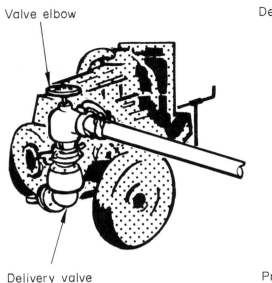

Valve elbow

Delivery valve

(a)

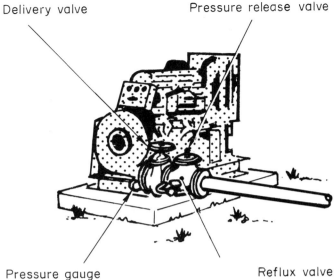
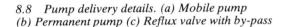

Delivery valve

Pressure release valve

Pressure gauge

Reflux valve

(b)

*8.8 Pump delivery details. (a) Mobile pump
(b) Permanent pump (c) Reflux valve with by-pass*

regulate the pressure in the sprinkler system when opening the delivery valve.

On mobile pumps the delivery valve sometimes also acts as a reflux valve. The valve is very similar to that used in portable irrigation systems (see figure 3.23) but the sealing disc slides freely instead of being attached to a screwed shaft. The flow holds the valve seal open. The amount of opening is controlled by turning the hand wheel. When the pump stops the sealing disc drops back on to its seating and prevents water flowing back into the pump. An elbow fits over the valve to connect with the mainline.

On permanent pumps the valves are normally separate. Some reflux valves are fitted with a small by-pass valve. This allows flow to pass around the valve and so can be used as a priming device.

8.5 Operation
Before starting a pump it is essential to check that:
1 All joints on the suction side are airtight.

2 All moving parts on the pump and power unit have been lubricated as recommended by the manufacturer.

3 For mobile tractor mounted pumps, the tractor and pump are correctly positioned and aligned.

4 All engine and shaft safety guards are in place.

5 For internal combustion engines, there is sufficient fuel available.

Priming

1 Close the delivery valve.

2 Prime the pump so that both pump and suction pipe are filled with water.

Starting

1 Follow power unit starting procedures normally provided by the manufacturer.

2 Pump pressure gauge should show a steady rise. If not try operating the priming pump for a few strokes if one is fitted.

3 If the pressure still fails to rise, stop the pump and repeat priming and starting procedures.

4 If pump fails repeatedly, stop the pump and find the cause of the problem (see section 8.6).

5 *Always stop a pump immediately if it fails to deliver water.* Parts of a pump are lubricated by water and the pump must never be allowed to run dry.

6 Once the pump is running properly, open the delivery valve slowly. If the valve is opened too quickly it may cause:

Loss of prime by allowing air from the mainline back into the pump.

Large pump discharges as the mainline and laterals fill and pressures build up. This greatly increases the power demand and may overload and damage the power unit.

Water hammer in the mainline and laterals because of the rapid changes in the flow.

7 When the sprinklers start to operate, this indicates that all the pipes are filled. The delivery valve can then be safely opened up. The valve can be set to give the desired operating pressure in the system.

Stopping

1 Before stopping the pump close the delivery valve slowly. This will avoid water hammer in the pipe system. Closing the valve will also help to retain the pump prime.

2 Stop the pump as soon as possible after closing the delivery valve.

3 Follow the power unit stopping instructions normally provided by the manufacturer.

For pumps driven by an internal combustion engine it is normal practice to open the throttle until it is running at only half the normal pump speed when starting. Once the operator is sure that everything is working properly the throttle is opened to the normal pump speed. Such a practice is not possible with an electric motor as it will only run at one of the standard speeds.

8.6 Common faults

When a centrifugal pump fails to operate or the pressure or discharge falls during operation, the pump should be shut down as quickly as possible and the cause of trouble investigated. Experience has shown that the majority of problems can be traced to faults in the suction side of the pump such as air leaks causing loss of prime and pipe blockages. The following are the most common problems and their causes:

Pump fails to prime

1 The most likely cause is air leaking into the suction pipe or pump. This may be due to:

Insufficient priming.

Leaks in suction pipe couplings.

Delivery valve not closed properly.

Suction pipe inlet not far enough below the water surface.

2 The suction pipe inlet may be blocked by debris.

3 Priming system may be faulty. When a priming pump is used check that no stones or weeds are fouling the openings, and ensure that seals are undamaged.

Pump fails to deliver sufficient pressure or discharge

1 Air may be leaking into the suction pipe or pump but is not enough to cause loss of prime.

2 Suction pipe inlet or pump impeller may be partially blocked with debris.

3 The suction lift may be too high.

4 Pump speed may be too low. A loss of 20% in pump speed will reduce the discharge by 20% and the pressure by 44%.

5 Pump bearing may be worn and water is leaking back into the suction side.

6 Pipes or line filters may be blocked. In such cases the pressure is increased but the discharge decreases. This is characteristic of a centrifugal pump running at constant speed.

Pump delivers water but later stops delivering

1 Many of the causes are similar to those already described above.

2 Air may have accumulated in a high spot in the suction pipe. If this suddenly moves up into the pump it may cause loss of prime.

3 Air may enter the suction inlet if the water level drops during pumping. Watch the water surface to see if there is a whirlpool near the inlet through which air can enter.

Pump takes too much power

1 There may be leaks in the mainline or laterals or a pipe burst. This causes the pressure to fall and the discharge to rise thus increasing the power required.

2 Pump speed may be too high.

3 There may be mechanical defects in pump or power unit. In such cases examine the lubrication system and check that the pump can be easily rotated by hand. If this is not the problem, open the pump and examine the impeller for wear or damage.

8.7 Care and maintenance

As with all machines, regular maintenance is essential if a pump is to provide reliable and efficient service:

1 Ensure that all pump components are lubricated as recommended by the manufacturer. Many pumps have water-lubricated seals. These seals must not be lubricated with oil.

2 Do not expect a pump to deliver water at a higher rate or pressure than it is designed for, eg by adding more sprinklers or pipes. Such overloading will cause excessive wear on the pump and shorten its working life.

3 Check the working pressures of a pump regularly against values provided by the manufacturer when it was new. This is most easily done when the delivery valve is closed and there is no flow. Make sure that the test is carried out at the proper pump speed.

At the end of an irrigation season.

1 Drain all water from the pump. This will prevent corrosion and any possible frost damage. There is often a small plug in the bottom of the casing for this purpose. Always open the delivery valve when draining.

2 Always leave the delivery valve open slightly when not in use to avoid the rubber seals from sticking to the seating.

3 If water seals are exposed during storage apply a little oil to the metal surfaces by rotating the impeller by hand (figure 8.9).

4 For portable pumps cover the inlet and outlet holes to prevent debris and rodents from entering (figure 8.9).

5 At the beginning of the next season always rotate the impeller by hand before starting up. The pump may have seized up while not in use and starting up may damage both the pump and the power unit.

8.8 Performance

A centrifugal pump is designed to run at a constant speed. Its performance can be described by:

Pressure and discharge

Power requirement

Efficiency

Pressure and discharge

A pump will deliver a wide range of discharges depending on the pressure required and the speed at which it rotates.

When a pump starts up it takes a little time for it to reach its normal running speed. During this time the delivery valve is closed and there is no flow in the mainline. Pressure gradually builds up inside the casing. A common fault is to think that because the delivery valve is closed, the pressure will go on rising until eventually it bursts the pump casing. This is not so with a centrifugal pump. The pressure depends only on the speed of the pump. The faster it rotates, the higher the pressure it produces.

8.9   *Care and maintenance (a) Applying oil to water seal (b) Covering inlet and outlet for storage*

As the delivery valve is opened the pump starts to discharge into the mainline. The discharge available depends on the pressure required. If a high pressure is needed then only a small discharge will be available. If only a low pressure is needed then a much higher discharge will be available. This is shown as a graph in figure 8.10(a).

Higher pressures and discharges can be obtained from a pump by increasing its speed.

Power requirement

A pump needs power to rotate the impeller at a constant speed. This is supplied by the power unit. It is usually measured as Brake Horse Power (BHP). The amount of power required varies with the pump discharge (figure 8.10(b)). When the pump starts and there is no flow, only a small amount of power is needed to rotate the impeller and water in the pump. As the delivery valve is opened and the flow starts, the power required to turn the pump increases. More power is also required to increase the speed of a pump.

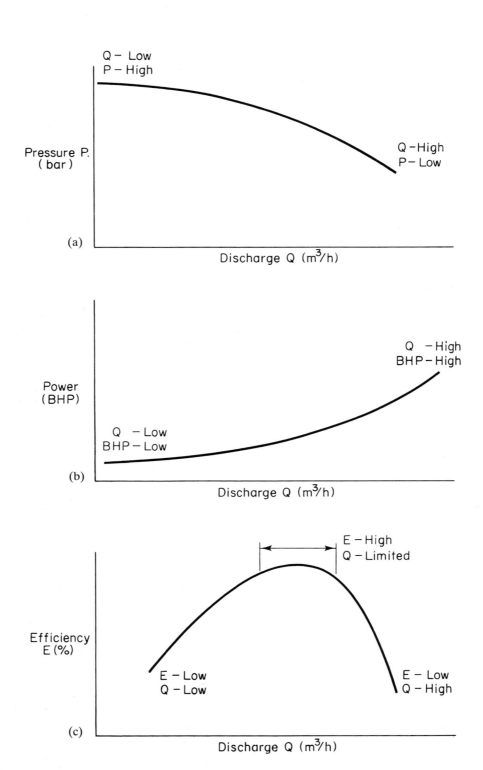

8.10 Pump performance. (a) Pressure and discharge
(b) Power (c) Efficiency

114

Efficiency

The efficiency of a pump measures how well the mechanical power supplied by the power unit is converted into water power in the pump. It is often written as:

$$\text{Efficiency} \quad \% = \frac{\text{Water Power produced} \times 100}{\text{Mechanical Power supplied}}$$

This expresses efficiency as a percentage (%).

Centrifugal pump efficiencies vary from 30% to 80% depending on how they are used. There is only a small range of discharges at which the pump will have the highest efficiency (figure 8.10(c)). If the pump operates above or below these values it will be less efficient and more power will be needed to operate the irrigation system.

All centrifugal pumps perform in this way but different pump sizes and shapes produce different pressures, discharges and power requirements. Information on pump performance is normally provided by the manufacturer. This helps the engineer to select a pump whose performance closely matches the needs of the irrigation system. Obviously the pump must be able to produce the required pressure and discharge but for best performance this should be near to the point of highest efficiency. In this way the power supplied will be used most efficiently.

8.9 Power units

There are two main types of power unit used to drive irrigation pumps:

Internal combustion engines

Electric motors

It is not possible to say that one type of unit will always be more suitable than the other for a particular location. It depends on many factors including:

Cost of installing, running and maintaining the power unit.

Whether installation is permanent or mobile.

Available power supplies.

Available labour skills to operate and maintain equipment.

Internal combustion engines

These are either spark ignited (petrol or gas) or compression ignited (diesel) and are suitable for both permanent and mobile pumps.

Petrol engines are not normally used in irrigation unless there is no other form of power available. Although they are relatively cheap to buy, they can be expensive to run using costly high quality fuels. They also have a relatively short working life and require a lot of maintenance.

Diesel engines are used extensively in both small and large irrigation schemes. In remote areas where no electricity supply is available they are usually the only form of power that can be used. Diesels are more expensive than petrol engines but their running costs are much lower. Provided they are operated and maintained properly they may have a useful working life of 15000 hours (approximately 15 years). They require constant supervision for fuelling and other routine tasks such as oil and filter changes. As there are a large number of moving parts regular engine overhauls are required. A top overhaul is normally required after 2000 hours operation and a major engine overhaul after 5000 hours. During its life a diesel may require three top overhauls and one major overhaul.

A modern diesel is usually fitted with safety devices so that it can be left for long periods without supervision (figure 8.11). These protect the engine when there is:

Lack of oil in the power unit or the oil pressure fails.

Lack of water in the pump.

Excessive engine temperature – through leaks in cooling system or overloading of the pump.

A burst on the mainline or laterals and the power suddenly increases.

A timer can also be used to shut down the engine after a pre-determined period of operation.

Diesels normally operate at only 60-70% of their maximum power output and should not be operated at full power for long periods. When irrigating, the power required by the pump can be controlled by slowly opening the pump delivery valve and by varying the engine speed using the throttle, to avoid excessive power demands.

Another source of diesel power commonly used in irrigation is from the power take off

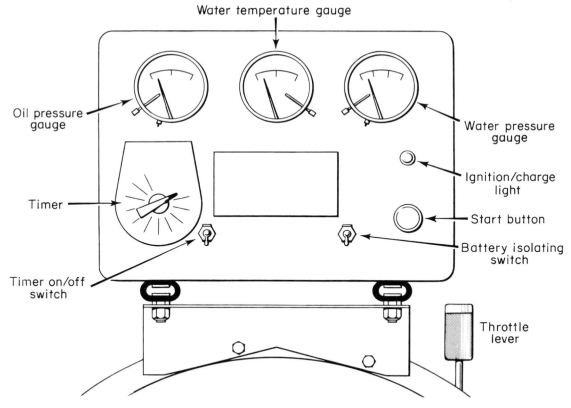

Water temperature gauge

Oil pressure gauge

Timer

Timer on/off switch

Water pressure gauge

Ignition/charge light

Start button

Battery isolating switch

Throttle lever

8.11   Safety devices on modern diesel engines

point (pto) on a tractor. Only 60% of the tractor power is available at the pto but this can be used to drive mobile centrifugal pumps. These are ideal for small jobs requiring intermittent pumping. If continuous pumping over long periods is required it may be better to use a separate diesel unit rather than tie up a tractor.

Electric motors

Electric motors are an alternative form of power to diesels when there is an electricity supply available. They are more reliable than diesel engines and can provide 20-30 years of almost trouble free life. They can operate for long periods unattended and require very little maintenance as there is only one moving part – the rotor. They are also simple to operate in terms of stopping and starting and this can be easily automated.

Electric motors are only suitable for permanent installations with long hours of pumping. They are not suitable for mobile pumps or for short period intermittent pumping. Electric motors work at one of a number of constant speeds which are shown in table 8.2, together with their power requirements.

Electricity supply		Common speeds (rpm)			
50 cycles per second (Hz)		725	960	1450*	2900
60 cycles per second (Hz)		870	1160	1750*	
		Power output of motor			
110v or 220v	Single phase	Up to 7 HP			
220v	Three phase	7 to 25 HP			
440v	Three phase	more than 25 HP			

*Most common electric motor speed for centrifugal pump use.

Table 8   Electric motors: speed and power requirements

The best arrangement is when the motor is close coupled to the pump. If different pump speeds are required a belt-drive can be used but this increases the maintenance needed.

Electric motors normally operate close to their maximum power output. Care should always be taken to open the pump delivery valve slowly so as not to increase the discharge and power demand on the motor. Unless this is done the motor may overheat and be shut down by a thermal overload switch before the required pressure is reached.

INDEX